COASTAL SEA
The Conservation Challenge

John R. Clark
MOTE MARINE LABORATORY
RAMROD KEY
FLORIDA, USA

b

**Blackwell
Science**

© 1998 by
Blackwell Science Ltd
Editorial Offices:
Osney Mead, Oxford OX2 0EL
25 John Street, London WC1N 2BL
23 Ainslie Place, Edinburgh EH3 6AJ
350 Main Street, Malden
  MA 02148 5018, USA
54 University Street, Carlton
  Victoria 3053, Australia
10, rue Casimir Delavigne
  75006 Paris, France

Other Editorial Offices:
Blackwell Wissenschafts-Verlag GmbH
Kurfürstendamm 57
10707 Berlin, Germany

Blackwell Science KK
MG Kodenmacho Building
7–10 Kodenmacho Nihombashi
Chuo-ku, Tokyo 104, Japan

The right of the Author to be
identified as the Author of this Work
has been asserted in accordance
with the Copyright, Designs and
Patents Act 1988.

First published 1998

Set by Excel Typesetters Co., Hong Kong
Printed and bound in Great Britain by
MPG Books Ltd, Bodmin, Cornwall

The Blackwell Science logo is a
trade mark of Blackwell Science Ltd,
registered at the United Kingdom
Trade Marks Registry

DISTRIBUTORS

Marston Book Services Ltd
PO Box 269
Abingdon, Oxon OX14 4YN
(*Orders*: Tel: 01235 465500
      Fax: 01235 465555)

USA
Blackwell Science, Inc.
Commerce Place
350 Main Street
Malden, MA 02148 5018
(*Orders*: Tel: 800 759 6102
            781 388 8250
      Fax: 781 388 8255)

Canada
Login Brothers Book Company
324 Saulteaux Crescent
Winnipeg, Manitoba R3J 3T2
(*Orders*: Tel: 204 224-4068)

Australia
Blackwell Science Pty Ltd
54 University Street
Carlton, Victoria 3053
(*Orders*: Tel: 3 9347 0300
      Fax: 3 9347 5001)

A catalogue record for this title
is available from the British Library

ISBN 0-632-04955-3

Library of Congress
Cataloging-in-publication Data

Clark, John R., 1927–
  Coastal seas: the conservation
challenge/John R. Clark.
      p.   cm.
  Includes bibliographical references and
index.
  ISBN 0-632-04955-3
  1. Coastal zone management.
2. Environmental protection.
3. Coastal ecology. I. Title.
HT391.C4935   1998
333.91′7—dc21   98-11892
                                        CIP

Contents

Preface

As our rapidly growing world approaches the 21st century and food and a clean environment become the basis for human welfare, it is important to combat pollution and conserve such valuable littoral resources as beaches, coral reefs, mangrove forests, and coastal lagoons that support fisheries and wildlife and filter pollutants from the water.

This book is about managing land and coastal water resources together in a program called unified "Coastal Zone Management" (CZM) which is tested and ready to help solve the problems of the 21st century. Unified, or "integrated", CZM recognizes that the coastal resources management situation is unique; that is, it differs greatly from either land or water resources management, being a combination of both.

Use of the term "Coastal Zone" indicates that the coast has been distinguished as a geographical zone apart from, yet between, the oceanic domain and the terrestrial domain. The near coast and littoral areas form the most bounteous part of the sea for both commerce and recreation. Shoreland, especially the waterfront, is an exceptionally scarce and valuable resource but its development threatens the health of coastal waters. Most countries have conservation programs that address land resources or that address water resources, but too few countries treat them together in a unified framework of CZM.

Our experience shows that CZM programs can indeed add to the economic and social prosperity of coastal communities. Enhanced fisheries productivity, increased tourism revenues, cleaner waters, and protection of lives and property from sea storms are among the practical benefits of coastal zone planning and management. To accomplish these things requires that governments, private business, and communities join together in a collective agenda for conservation.

This book in congruent with the modern conservation management concepts of sustainable use, multiple use, rights to the commons, biodiversity, zoning, protection of special habitats, public participation, capacity building, institutional strengthening, co-management, situation management, and integrated management. The use of modern computer and cybernetic utilities is emphasized.

The content deals with conservation methodology supported by a compatible philosophical base. The book is intended to satisfy a worldwide demand for a succinct technical reference on coastal zone resources management. The author attempts to provide the reader with the ideas, management formats, and tools and materials needed to manage coastal natural resources and shoreline developments. While it will be of interest to all 170 coastal countries, tropical developing nations have the greatest need for this book. It is crafted to be of maximum value to them. The book tries to enlighten but not persuade, and to coach but not prescribe.

The author expects to find a readership among professors, scholars, scientists, policy analysts, students, consulting companies, government conservation officers, management authorities, legislators, coastal developers, design engineers, international development agencies (e.g. USAID, UNDP), private conservation groups, research institutes, and libraries (general, special, and university), and also a variety of supporting professionals such as lawyers, sociologists, biologists, geologists, geographers, and planners.

While this little book presents knowledge garnered from the work of numerous colleagues around the world, the responsibility for accuracy and relevance is strictly the author's. The book is a very brief review, intended to give the reader a toehold on the extensive literature dealing with CZM. It provides a starting point. CZM is dealt with in much greater detail in my *Coastal Zone Management Handbook* [1].

Many particulars can be found in the references that are cited and listed in the References section. Especially noteworthy regarding tropical areas are the works of Sorensen, Burbridge, Maragos, Kenchington, Thia-Eng, Olsen, Saenger, Hotta and Dutton, Bacon, White, Cambers, Salm, Tomasik, and Wells. I have not given sources for items of common knowledge.

Why did I write this book? Because someone had to. A critic once called Bacon's *Organon* "A book which a fool could not and a wise man would not have written." I hope that I too fall in the middle.

Acknowledgements

The people who helped with this book form a nearly endless list. Their help extends back as far as 15 years because this book is part of a stream of publication starting in 1982. Numerous colleagues assisted me with this book, and none ever asked why I was writing it or why it was needed. Still I did sense a certain air of resignation among those who I called upon for the fourth or fifth time. These helpers are credited at the place of their contributions.

Colleagues who deserve special mention are Jens Sorensen, University of Rhode Island; Alan White, PRC Environmental Management, Cebu, Philippines; Peter Burbridge, Newcastle University, Newcastle, England; Peter Saenger, Southern Cross University, NSW, Australia; Alec Dawson Shepherd, Hunting Associates, York, England; Wim Verheugt, Euroconsult, Arnhem, the Netherlands; Kay Hale, Library Director at the Rosensteil Marine and Atmospheric School of the University of Miami; and my friend and computer expert, Edgar Piehl of Ramrod Key, Florida.

I am especially grateful for the diligence, enthusiasm and charm with which my editor at Blackwell Science, Ian Sherman, pursued this project from the very beginning and for the masterful handling of the manuscript by production editor Bridgette Jones.

I am also most grateful for publisher permissions to use material from FAO's *Integrated Management of Coastal Zones* and CRC/Lewis's *Coastal Zone Management Handbook* along with graphics from various sources, as credited in figure titles.

1: Viewpoint

*Now the great winds shorewards blow; now the salt tides
seawards flow.* [Matthew Arnold, 1849]

The shoreline is a frontier that stands between two worlds—the solid world
and the liquid world. Both worlds, earth and water, hold resources of
immense value to mankind. But most of these resources are at risk of depletion and urgently require conservation management. Such management
must embrace the land side of the "Coastal Zone" as well as the water side.

Because the Coastal Zone straddles the coastline and spans the transition
from sea to land, it has many features unique to the coast like daily tides,
mangrove forests, coral reefs, tidal flats, sea beaches, storm waves, estuaries,
and barrier islands. The Coastal Zone is not only distinctive, but extremely
productive of renewable resources such as protein food, tourist income,
mangrove products, and other economic goods and services. But conservation of these resources often clashes with shorelands development.

This chapter presents the author's viewpoint on concepts for conserving
coastal natural resources and their genetic basis. It lays the groundwork
for assimilating the ideas and information provided in the next eight chap-
ters, in keeping with the new conservation concept for the 21st century—the
"eco-realism" of Easterbrook, which encourages community self-interest,
enlightened by the inevitability of natural law [2].

Problems. The Coastal Zone is a major attraction to ocean commerce,
tourism, home seekers, the military, and a variety of industries. In the shorelands, dense populations are attracted. At the shoreline (the "littoral") great
environmental modification is caused by land conversion, sea dredging, and
water pollution from urban, industrial, commercial, and agricultural development. In the more offshore waters, oil exploitation, ocean dumping, min-
erals mining, and excessive fish harvesting are potential threats.

Coastal zones are the "sink" for the pollution from the land side. They
receive and concentrate pollutants and suffer other negative impacts of activities taking place in the shorelands and hinterlands that deplete coastal
resources. Conversely, the sea strongly affects the land and intertidal areas;
for example, beach erosion, pollution from tanker bilge washings, and
property destruction from cyclonic storm surges, flooding, and wave action.

Overall, depletion of coastal resources is the result of ignorance, careless
development, and overexploitation by both the private sector, public agencies and communities. The depletion is aggravated by development with
short-term profit horizons and by the absence of defined user rights.

Really heavy damage to coastal resources started in the 1960s and accelerated through the 1970s. In the 1980s and 1990s worldwide awareness of the need for conservation was achieved. We hope that in the 21st century this awareness will stimulate serious conservation programs in all countries.

Behind the devastating coastal resources depletion suffered by many countries are several major driving forces, such as the following (paraphrased from [3]):

1   high rates of population growth;

2   poverty exacerbated by dwindling resources, degraded fisheries habitats and lack of alternative livelihoods;

3   large-scale, quick-profit enterprises which degrade resources and conflict with interests of coastal communities;

4   lack of awareness about management needs for resource sustainability among coastal people and policy makers;

5   lack of understanding of the economic contribution of coastal resources to society; and

6   lack of serious government follow-up in support and enforcement of conservation programs.

Coastal Zone Management. The lesson to be learned is that, in order to ensure the maximum quantity and quality of renewable natural resources for ourselves and our descendants, we must learn to use resources sustainably. To do this we must secure the cooperation of communities, industries, and governments in conservation campaigns which, for the coast, can be done effectively through a Coastal Zone Management (CZM) program.

CZM is a distinct process which deals with distinct forms of geology, climate, biota, hydrology and natural habitats. Demographic, socio-cultural, and political systems are also distinct in many ways—the coastal location typically creates distinctively maritime cultures. Planners, managers, and politicians must understand that because of these distinctions special programs for coastal resource conservation are required.

CZM is meant to combine the management of coastal waters and shorelands. Many countries have programs to conserve land resources and/or marine resources, but frequently do so separately. Too few countries combine "land side" and "water side" management in a single unified CZM framework.

CZM is a powerful mechanism for allocation of natural resources and control of bad development when it is based upon sound environmental and socio-economic planning and evaluation. It requires networking among all relevant government activities, including national economic development planning. The key is *unitary management* of the zone, which treats the shorelands and coastal waters as a single interacting unit and coordinates the interest of all stakeholders (those with interests in the Coastal Zone) with a collective agenda.

A primary strategy of CZM is to regulate construction and other actions in the coastal zone, often through a project review and permit letting process. CZM attempts to guide *future development* as a main purpose while also trying to correct environmental mistakes of the past as a parallel purpose.

The use of a particular coastal resource for a single economic purpose is discouraged by CZM in favor of a balance of multiple uses whereby economic and social benefits are jointly maximized and conservation and development become compatible goals (with the term "conservation" used in its classic sense). The demand for coastal space and resources is usually so great that no one activity can be given exclusive use. Many uses can coexist in a CZM program that encourages multiple use, while others may have to be restricted.

Some countries have organized full scale CZM programs (e.g. the United States, Sri Lanka) or partial programs (e.g. Philippines, Australia, Costa Rica) while many others are evaluating unified CZM programs. To date 56 countries are said to have tested various approaches to CZM [4].

The Commons. In most countries coastal waters and their resources are considered "commons"; that is, they are not owned by any person or agency but are common property available equally to all citizens, with the government as "trustee" — this is an age-old public right, *jure communia,* going back to the Institutes of Justinian: "*Et quidem naturali jure communia sunt omnium haec: aer, aqua profluens, et mare per hoc litora maris*". In English this means: "By the law of nature these things are common to mankind: the air, running water, the sea and consequently the shores of the sea".

Further, this influential doctrine states that: "No one therefore is forbidden to approach the seashore, provided that he respects habitations, monuments, and buildings, which are not, like the sea, subject only to the law of nations". And now "environment" must be added to the "he respects" list. A primary aim of CZM is to provide for sustainable use of the resources of the Commons, a responsibility that should be shared by all people and all levels of government.

Boundaries. There is no single definition of the Coastal Zone. Its boundaries are delineated on the basis of the particular problems that the management program is supposed to solve. Because CZM is an "issue based" approach, the boundaries must adapt to the defined issues and to program objectives. Because of the linkages between the land side and the water side of the coast, CZM boundaries are set to encompass a zone that includes both shorelands and coastal waters.

By any set of criteria, the coastal zone is a linear band of land and water that straddles the coast—a "corridor" in planning parlance. This corridor has a one-dimensional aspect; the second dimension (width from onshore to offshore) tends to be overshadowed by the linearity. This is reflected in the common parlance; i.e. people talk about being *at* the coast or *on* the coast, but never *in* the coast.

Conservation. The word "conservation" (in its English language version) came into use early in the 20th century in the United States during the administration (1901–8) of President Theodore Roosevelt to mean, simply, "wise use" of natural resources or one might say, the control of exploitation to ensure resource sustainability, which is the alternative to resource exploitation for short-term profit.

The criterion for sustainable use is that the resource not be harvested, extracted or utilized in excess of the amount which can be simultaneously regenerated. In essence, the resource is seen as a capital investment with an annual yield; it is therefore the yield that is utilized and not the capital investment which is the resource base. Planners should view the Coastal Zone as an economic entity, like farmlands, production forests, rangelands, or cities.

In the classic sense "conservation" and "sustainable use" are nearly analogous and sustainable use has become the modern catchword for classic conservation. Unfortunately, in a broader vernacular use, conservation is now often used interchangeably with such words as "protection" and "preservation", and "conservationist" is used interchangeably with "environmentalist" or "ecologist".

Environmental conservation enters into the management equation in two major senses: (i) *material*, ensuring the sustainability of economic resources, which may be termed "conservation"; and (ii) *spiritual*, including the important, less tangible, values of nature protection and biodiversity conservation, which may be termed "environmental protection". Inclusion of spiritual resources in a unified CZM program is seen as a luxury in many countries which tend to give the priority to the material side—tangible yields, products, and consumption.

2: Coastal Resources Status

Lawfull to you is the pursuit of water-game and its use for food—for the benefit of yourselves and those who travel.
[The Koran, Sura 5, aya 96]

The Coastal Zone supports a large part of the world's living marine resources, certainly more than the open sea. This zone also has the highest biological diversity of any part of the sea. Its "Special Habitats" provide breeding and feeding places and nurture the young of many species. The characteristics and values of these Special Habitats are discussed in this section.

With conservation of Special Habitats as a management priority, it is necessary to reckon their vulnerabilities and conservation needs. While the whole of a coastal ecosystem is the focus of management, each of its structural parts and essential processes must be individually conserved — mangrove forests, tide-flats, beaches, seagrass or kelp beds, and coral reefs. For management, one can look at the coast as a collection of habitat types united by water flow.

2.1 *Mangrove wetlands*

The term "mangrove" refers to any of dozens of coastal species of trees capable of living in saltwater or salty soil regimes. Mangrove forests are a major resource of tropical and subtropical sheltered coastlines where wave activity tends to be moderate. They are biologically rich and contain a great number of plants and animals (e.g. crustaceans, molluscs, fishes, and birds) which often form a significant part of the coastal resource. In their natural state they are very prolific, e.g. in the Sundarbans of Bangladesh (7800 km²), seedlings are produced at the rate of 27 750 hectares per year [5].

Ecologically, a prominent role for mangrove is the production of leaf litter and detrital matter which is "exported" on the tide to lagoons and the nearshore coastal environment as a rich nutrient for sea life. Shrimps and many fish species are supported by this food source. Also, this habitat is directly utilized as a nurturing place for the young of species from other nearby habitats: estuaries/lagoons, near-coast waters, seagrass meadows, and coral reefs. A variety of species of birds, mammals, and reptiles find shelter in the mangrove forest.

Moreover, mangrove forests enhance water quality by extracting chemical pollutants from the water (in moderate quantity). Shoreline mangroves are recognized for their ability to stabilize coastal shorelines that would otherwise be subject to erosion and act as a buffer against cyclonic storm-tide surges that would otherwise wash more strongly over low-lying areas and threaten life and property.

Mangrove forests provide life support and income for millions of people for wood, charcoal, honey, and extractives. All these "free" services are an economic boon to adjacent communities. Nevertheless, mangrove forests have been obliterated in dozens of countries in order to: (i) create real estate by land filling; (ii) make shrimp ponds by excavation; (iii) produce charcoal and wood by clear felling; and (iv) build ports and other infrastructure. Many countries have created special mangrove conservation programs to stop rampant exploitation of the world's remaining 24 million hectares of mangrove forest. But where sufficient restrictive laws have been passed, enforcement is often not strict enough.

In general, the mangrove ecosystem is robust. But mangroves can be affected by excessive sedimentation, stagnation, and major spills of pollutants (including oil). These actions, which reduce the uptake of oxygen for respiration, can result in rapid mortality. Salinities high enough to disable mangroves (+90 ppt) may result from reduction of freshwater inflow (e.g. by dams). Restriction of tidal flow can also kill mangroves.

2.2 *Fringing intertidal systems*

Intertidal areas of the coastal edge are sometimes covered with grasses and rushes rather than mangrove trees. Such "salt marsh" wetlands, where they exist, serve many of the same ecological purposes as mangrove forests and are classed as Special Habitats. For example, they too assimilate nutrients and convert them to plant tissue which is broken into fine particles and swept into the coastal waters as plant detritus. Salt marshes, among the most productive ecosystems in the world, may produce up to 10 tons of biomass annually. In terms of wildlife productivity, the salt marshes play a major role by contributing food, shelter, and nesting sites for thousands of waterfowl, alligators, muskrats, crabs, and shrimp, and by nurturing the young of coastal species [6].

Salt marshes have suffered greatly from development activities. Filling, dredging, ditching, impounding, and draining, as well as polluting, have greatly reduced the total acreage. Ditching has a profound effect in reducing invertebrate productivity and modifying the vegetational pattern. Impounding destroys the salt-marsh vegetation. Causeways constructed for highways and railroads restrict tidal flow as do tidal gates.

In addition to marshlands and mangroves, coastal zones often have wide areas of open tide-flats (mud-flats, sand-flats, etc.), often found in estuaries and lagoons. Such flats are important in providing feeding areas for fish at high tide or birds at low tide. Mud flats are often important energy processing and storage elements of lagoon, estuary, and delta ecosystems. The tideflat serves to catch nutrients and hold them. In many littoral basins, tide-flats produce a high yield of shellfish.

2.3 Seagrass meadows

Submerged seagrasses are often abundant in the shallow waters of the Coastal Zone. Classed as Special Habitats, seagrass bottoms are highly productive and valuable assets. Seagrasses grow best in the quiet, protected waters of healthy estuaries and lagoons — often in beds, or meadows — in waters clear enough that light can penetrate to enable photosynthesis to occur. Productivity of seagrass meadows is enhanced by the secondary small plants, or "epiphytes", that attach to the leaf surfaces. In fact, primary productivity (amount of plant production) for two common seagrass species has been shown (USA) to be higher per acre than ordinary corn and rice fields.

Wherever they do occur, seagrass meadows are essential elements of coastal ecosystems. These meadows provide vital places of refuge not found in the open ocean. They attract a diverse and prolific biota and serve as essential nurturing areas for some important marine species. Seagrass meadows are also known to trap and bind sediments, thereby reducing particulate pollutants. They also export large amounts of detrital nutrient.

However, seagrass meadows are being depleted by: (i) widespread dredge and fill activities; (ii) by water pollution, including brine disposal from desalination facilities, waste disposal from factories, spills of petroleum and petroleum products; (iii) thermal discharges from power plants; (iv) excessive siltation and turbidity; (v) bottom trawls which scrape and plough the meadows; and (vi) marine excavation and filling. But because so much of the damage is unseen, it is too often overlooked.

2.4 Kelp systems

At the higher latitudes, e.g. south-west South Africa and California, submerged forests of kelp form Special Habitats. Kelp is a brown alga that roots on the bottom of the sea and extends, via a long stalk, to the surface where its fronds spread out over the sea surface. Kelp often grows in thick stands that provide an important habitat from surface to bottom for many valuable species of fish, shellfish, and mammals (e.g. sea otter). But these kelp stands are jeopardized by: (i) overharvesting (for alginic acid); (ii) species imbalance (e.g. too many kelp-eating sea urchins); and (iii) pollution and other impacts.

2.5 Coral reef systems

Coral reefs occur along shallow, tropical coastlines where the marine waters

Fig. 1 Coral reefs provide habitats for a broad range of species. (Photo by R. Stone.)

are clean, clear, and warm. They form one of the most productive ecosystems in the world because of both the productivity of corals and collaboration of the coral reef ecosystem with its surrounding environment (Fig. 1).

Coral reefs form in many configurations. "Barrier" reefs are offshore linear reef structures that run parallel to coastlines and arise from submerged shelf platforms. The water area between the reef and the shoreline is often termed a "moat" or "coral lagoon". The areas of greatest coral reef development are the Western Pacific and Indian Ocean and, to a lesser extent, the Caribbean Sea, including Belize and the Bahamas. The world's largest barrier reef system, the Great Barrier Reef, occurs off the Queensland coast of Australia.

"Fringing" reef systems—those contiguous with the shore—are the most common and widespread of the reef structural types; they are most usually found below the low-tide level. Some fringing reefs have a quite shallow moat between the reef crest and the shoreline. Their inshore location makes fringing reefs especially susceptible to degradation from coastal activities. One of the largest fringing reef formations lies along much of the Saudi Arabian Red Sea Coast (1770.2 kilometers) which is noted for an absence of freshwater (and sediment) runoff [7].

"Patch" reefs are isolated and discontinuous bits of coral reef, usually lying shoreward of barrier reefs in the coral lagoon, or moat. There are other reef types, such as the often circular "atoll" type which is common in the Western Pacific.

Coral reefs have important economic outputs. For example, they contribute to fisheries of three different types: (i) fishing directly on the reef; (ii) fishing in shallow coastal waters where coral reefs support food webs and life cycles; and (iii) fishing in offshore waters where the reef's natural productivity may be exported to help support adjacent deep sea fishes. About one third of the world's fish species are said to live on coral reefs [8]. Artisanal fisheries dependent on coral reefs are said to account for up to 90% of the fish production in Indonesia and up to 55% of production in the Philippines (providing 54% of the protein intake of all Filipinos).

In many countries, coral reefs support booming tourist industries that cater to snorkelers, divers, underwater photographers, sightseers, and sport fishermen. Reef-supported tourism produces billions of dollars of foreign exchange earnings annually. As an example, more than half of the foreign exchange earnings of the Cayman Islands are from coral reef based tourism.

In many coastal regions, coral reefs provide a great service as natural protective barriers, retarding storm waves, deterring beach erosion, protecting shoreline properties, allowing mangroves to prosper, and providing safe landing sites for boats.

The beachfront protected by coral reef is often the best place for dwellings in tropical countries. Unfortunately, there are numerous destructive forces at work and important coral resources are being degraded throughout much of the tropical world.

In many countries, reefs have been heavily exploited for corals which are harvested for sale as souvenirs or decorations; the market for coral is often quite lucrative and usually export oriented. Other damaging activities include: (i) mining of coral for construction materials; (ii) siltation and sedimentation created by dredging, filling, and related construction activities and increased soil erosion; (iii) pollutants, including spilled oil, industrial wastewater, and domestic sewage; (iv) a periodic discharge of large volumes of fresh water as may result from diversions and storm-water outfalls; (v) destructive fishing practices, including dynamiting; (vi) taking young fishes for sale in the aquarium trade; and (vii) tourist visits to reefs which result in breakage from boat anchors and from hand and foot damage.

In addition, there are damages from natural causes such as: (i) outbreaks of reef-destroying animals such as the crown-of-thorns starfish; (ii) diseases like whiteband (which kills elkhorn coral) and blackband (which kills large structural corals); (iii) hurricanes that smash the coral and sand blast away the living tissue; (iv) coral disablement and death from "bleaching" episodes (probably from high sea temperatures); and (v) die-off or depletion of essential symbionts, such as parrot fish and sea urchins that clean the reef of algae.

Some reefs are virtually beyond repair (those closest to settlements) but many that are degraded could still be returned to good or fair condition. For example, protect its coral reefs Sri Lanka created a unified CZM program in the early 1980s aimed especially at constraints on the mining of coral (see Sri Lanka case, Section 8.15).

2.6 *Beach systems*

Ecologically, the beach is a unique environment occupied by animals that have adapted to the constant motion of sand, gravel, or shell. Beaches provide a unique habitat for burrowing species such as mole crabs, coquina clams, razor clams, and crustaceans. Many important birds, reptiles, and other animals nest and breed on the dune, berm or open beach, as well as feed and rest there. For example, sea turtles come ashore during the spring and summer to lay their eggs in the "dry beach" above the high-water line. Also, terns and other seabirds frequently lay their eggs on the upper beach or in the dunes [9].

Serious socio-economic problems may arise with erosion and loss of beaches (e.g. risk to property, loss of international tourism revenue). On a worldwide basis, erosion (natural and human-induced) dominates over deposition, partly because of the global rise in sea level which is expected to accelerate (current rate of about 0.5–1.0 meter vertical rise per century in many parts of the world) owing to the "greenhouse effect". No tropical region seems totally free of the beach erosion problem (see Sections 8.1 and 8.11). Much of this is because of the occupation of beachfronts by residences and commercial structures and because of the "boomerang" effect of bulkheads, seawalls, jetties, and other constructions (see Sections 3.13 & 4.7.10).

2.7 *Lagoon and estuary systems*

Confined littoral basins—lagoons and estuaries—are an important component of Coastal Zone Management (CZM) programs. These basins are semi-enclosed, sheltered, shallow water bodies that have special ecological characteristics and provide essential natural services. These also include small bays and sounds, harbors, esteros, and so forth.

Natural lagoons are basins protected from the force of waves and which receive little fresh water and are therefore salt water influenced with relatively high salinity. These basins are sometimes formed by land configuration or sediment deposits but also exist often behind coral reefs or barrier islands. Lagoons exist widely, occupying 15% or more of the world's coastline.

Estuaries are brackish basins supplied by fresh water from rivers. They exist widely in the tropics except for arid or semiarid regions where major rivers are few and discharge is sporadic. Areas of particularly extensive estuarine environment include Brazil, West Africa, the Bay of Bengal shores of India and Bangladesh, and the Atlantic and Gulf of Mexico coasts of the United States.

Both estuaries and lagoons maintain exceptionally high levels of biological productivity and play important ecological roles such as: (i) "exporting" nutrients and organic materials to outside waters through tidal circulation; (ii) providing habitat for a number of commercially or recreationally valuable fish species; and (iii) serving the needs of migratory nearshore and oceanic species which require shallow, protected habitats for breeding and/or sanctuary for their young (nurture areas). To give one example, over 90% of all fish caught in the Gulf of Mexico are reported to be "estuarine dependent" to some degree [1].

Lagoons and estuaries are often the locus of substantial economic activity. Fisheries, shipping, commerce, industry, tourism, and housing crowd their shores. If this growth is not guided, the result can be rapid and severe degradation of the lagoon ecosystems and depletion of natural resources (see Sri Lanka case, Section 8.16). Many of these basins were diminished when large areas of lagoons and estuaries were reclaimed (drained and/or filled) to create ports or real estate or agricultural land, most notably in land-scarce regions (e.g. Japan and the Netherlands).

A big problem is pollution. Aside from outright fish and shellfish kills and other dramatic effects, pollution causes pervasive and continuous degradation, deterring mobile species from entering littoral basins. The most likely sources of pollution are agricultural and industrial chemicals and sewage and industrial organic wastes.

2.8 *Seafood resources*

Fisheries are essential to the economy and well being of citizens in numerous coastal countries. According to Brown [10], yearly fisheries harvest "... exceeds world beef production by a substantial margin and supplies ... 23% of all animal protein consumed in the world". For many countries, fish are the principal source of animal protein. Also, fisheries give jobs to millions of people—fishermen, boat builders, trap and net makers, packers, distributors, and retailers—all of which enhances social, cultural, economic, and political stability in the coastal areas. A strong domestic fishery promotes self-sufficiency, assists with the balance of trade, and reduces rural migration to cities.

3: Impacts

Do no mischief on the earth after it hath been set in order.
[The Koran, Sura 7, aya 56]

The Coastal Zone is a place of high-priority interest to people, commerce, the military, and a variety of industries. It is also a place that is easily damaged by careless development. The threats facing coastal zone natural resources as we enter the 21st century include land use intensification, over-exploitation of resources, preemption of natural areas, the effects of climate change, and many types of pollution.

Major coastal zone problems affect nearly 170 coastal states, but not all are equally affected. There are big variations in geology, climate, flora, fauna, water chemistry, sea condition, and littoral habitats from country to country. Equally great are the differences in resource uses, in demography, and in socio-cultural, economic and political systems. But there are also many commonalities.

Below we discuss various types of resource uses and abuses and their impacts on coastal resources and the security of coastal communities. The impacts of uncontrolled development must be understood and countered because of overexploitation, pollution, or alteration of natural systems, as well as conflicts between the users for limited, but highly prized, coastal and shoreline space. This chapter identifies several types of resource damage and ascribe them to their sources, namely, to particular "economic sectors".

3.1 *Origin of impacts*

The shoreline is a place where there is great competition and conflict between users and where governments need to develop the type of special policies and programs that unified Coastal Zone Management (CZM) offers. A large proportion of world population growth has occurred in coastal zones of developing countries; this has been accompanied by rapid changes in land and resource use. Increasing urbanization, industrial development, agricultural intensification, and expansion of tourism will all have profound effects on coastal habitats, and species.

Development planners must recognize, particularly, that modification of the land area (e.g. land clearing, filling, and grading) has a high potential for adverse effects on lagoon, estuarine, and nearshore coastal systems. Site preparation for development and expanded road networks, dams, industrial complexes, and other infrastructure are likely to further stress ecosystems and deplete coastal resources, as explained below.

3.2 *Typology of impacts*

The sources of environmental impact in coastal zones are varied and are categorized and defined in different ways by different sources. We make the following distinctions.

1   Direct impacts. Wetland clearing and filling, navigation dredging, coral extraction, overfishing, mangrove cutting, and waste discharges are all examples that have obvious immediate effects.

2   Indirect (or external) impacts. Inland terrestrial activities such as deforestation, improper soil coverage, improper building of infrastructure, and the indiscriminate use of agrichemicals all contribute to environmental degradation directly or indirectly.

3   Cumulative impacts. Often a series of minor impacts add up to a significant effect on coastal resources.

4   Ecosystem impacts. These may be a result of any of the above categories but pertain to a broader scale of changes with more long-term effects and the possibility of permanent alterations to whole ecosystems.

5   Socio-economic impacts. These result from direct or indirect environmental changes which impact the economic condition of communities or effect the socio-cultural life of the people.

6   Naturally caused impacts. These derive from natural forces such as global temperature changes, El Niño, hurricanes (cyclones, typhoons), floods, diseases, and conditions which result in the gradual or sudden death of organisms or in toxic effects on humans.

3.3 *Urbanization*

Coastal land use can strongly affect the sea. Impacts on coastal ecosystems arise from: industrial and agricultural pollution; filling to provide sites for industry, housing, recreation, airports, and farmland; dredging to create, deepen, and expand harbors; the excessive harvesting of mangroves, and so forth. The impacts reduce biological diversity, abundance of natural resources abundance (food and fiber), quality of life, community security (from sea storms), and tourism revenues. Although impacts are an unavoidable consequence of urbanization and human progress, they often can be greatly reduced by appropriate planning and management. Unified CZM anticipates such effects and offers solutions, particularly to careless development and the *unnecessary* sources of damage that result.

Human settlements can crowd the seas with boats, overexploit resources, generate polluting industries, produce heaps of garbage, and create large amounts of chemical wastes. The density of urban population creates demand for factories, power plants, warehouses, homes, docks, streets, automobiles, water supplies, and waste disposal. It results in competition with the needs of Nature, threatens the system's integrity, and impinges on human sensibilities by reducing natural resources and biodiversity and by creating ugliness.

While waterfront expansion may be economically advantageous for coastal cities, it can be threatening to coastal resources if development interests do not respect conservation principles. The Special Habitats of the littoral, e.g. mangrove wetlands, tide-flats, and seagrass beds, are too often obliterated when settlements preempt the shorelines, wetlands, and shallow seas by filling for real estate such as was done at Kuala Muda, Malaysia, where one massive project was said to create US$11.7 billion of real estate from "reclaimed" sea coast [11].

In many of the more densely populated nations, the risks of natural disasters to inhabitants of coastal lowlands are rising because of population increases, migrations to the coast, and poorly planned development projects. Coastal people become more susceptible to natural hazards such as floods, typhoons, or tsunamis when land reclamation projects encourage settlement in dangerously low-lying areas, or when land clearing and construction removes protective vegetation, reefs, or sand-dunes. A particularly disastrous example is Bangladesh where hundreds of thousands of people have been killed in recent major cyclonic sea storms.

Sewage treatment, water supply, electric power production, and impacts of other utilities and services add stress to resources and biodiversity. Electric power plants take in huge amounts of seawater for cooling purposes often killing masses of fish and planktonic organisms in the water [6]. The persistence of pesticides and other chemicals in littoral waters is of great concern. These can have a direct impact on the suitability of fish for the human diet.

3.4 *Waste disposal*

Currently, about 50% of the world's urban wastewater is directly discharged into the sea or nearby water bodies. Sewage and industrial outfalls, rivers, land runoff, and the atmosphere are believed to be responsible for most marine pollution. The rest comes from shipping, dumping, and offshore mining and oil production.

Polluted water is unhealthy for humans and for marine biota. Aside from outright fish kills, red tides, cholera outbreaks, and other dramatic effects, pollution causes pervasive and continuous degradation that is evidenced by the gradual disappearance of fish or shellfish or by a general decline in productivity. Ecological damage is concentrated in the lagoons/estuaries, and shallow littoral waters of the open coast.

Coastal communities often dump large amounts of solid wastes onto the Coastal Zone which disfigure the coastal landscape and leach pollutants into the coastal seas. Also, concentrations of septic tanks or cesspools leach large amounts of nutrient into shallow waters. Sewage outfalls (discharge) are particularly troublesome when inappropriately sited or if the waste is inade-

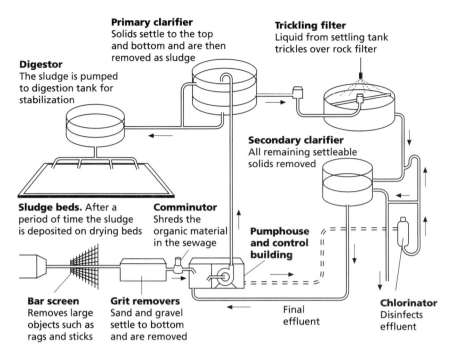

Fig. 2 Components and processes of a typical secondary sewage treatment plant. (Source: NACRF [12].)

quately treated. They can overload the marine environment with nutrients and discharge pathogens and toxic matter (Fig. 2). Coral reefs are particularly susceptible to wastewater and are easily upset ecologically (see Hawaii case, Section 8.19). Research in Barbados showed that nutrients can be beneficial to coral ecosystems in small amounts but destructive in larger amounts [13].

With intensifying demographic pressures, the sheer volume of waste products will increase. The need to protect human health provides a strong incentive to avoid discharges of pollutants into the marine environment.

3.5 *Agriculture*

Agriculture has strong implications for coastal conservation. Land clearing and farming often cause massive erosion of soil which ends up in coastal waters. Chemical applications, without careful control, run down rivers to the sea and affect water quality, contribute towards the eutrophication of river and coastal waters, enter into food chains, and debilitate ecosystems. Conversion of natural coastal lands to rice farming and other agriculture may preempt ecologically valuable wetland habitats. Lowlands conversion exacerbates flood hazards which not only destroy property and risk lives, but also drown crops and overwhelm estuary or lagoon ecosystems with sediment and organic and chemical pollutants.

3.6 *Forestry industries*

Excessive exploitation of forests for lumber, farmland, plantations, and fuelwood needs is a common practice. Massive deforestation not only leads to the depletion of a valuable resource, but also reduces watershed protection, increases surface water runoff, increases soil erosion and desertification, contributes to the siltation of dams, and results in pollution of the coastal zone. Clearly, uncontrolled removal of forest cover in watersheds increases sediment loadings of rivers and leads to smothering of estuarine organisms such as oyster beds, coral reefs, and seagrass meadows, which are Special Habitats of key importance for seafood production.

Reckless exploitation of mangrove forests, which has impoverished coastal people, is not so much the fault of organized forest industries as of land clearing interests (e.g. real estate, aquaculture, charcoal production). Some coastal communities rely heavily upon products from mangrove forests such as timber, tannins, fuelwood and charcoal, honey production, and sundry domestic products, without which they face poverty.

3.7 *Fishery industries*

The negative impact of fisheries on the coastal environment is usually slight but occasionally significant. The more environmentally damaging methods of fishing — dynamite fishing, chemical fishing, bottom trawling in shallow water — are prohibited by law in most countries, but enforcement is often weak. Unfortunately, excessive exploitation has depleted coastal fish stocks in most countries—not only are there too many fishermen, but their nets and traps are too often made so as to kill small fish needlessly. But fisheries are more often victim than perpetrator of environmental impacts.

A major global problem of the 21st century will be continuing depletion of seafood resources caused by coastal pollution and habitat destruction. To counteract this trend it is necessary to conserve the productivity of coastal habitats for fisheries, the same as we protect rangelands for livestock, farmlands for crops, and forest lands for wood. It is essential that seafood landing and processing facilities are ample. These are all important functions of unified CZM.

3.8 *Aquaculture*

Culture of marine species spreads across the boundary between land and sea; for example, shrimp ponds in mangrove wetlands or salmon cages and artificial reefs in littoral waters. Aquaculture is a fast growing sector in some countries as a valuable supplement to local diets or a means of earning foreign currency. Aquaculture may be both a polluter of the littoral and a petitioner for clean water (for its ponds and tanks).

Aquaculture can be harmful. For example, shrimp ponds often preempt ecologically valuable mangrove forest area. The effect may be to reduce natural reproduction of species (including those used in aquaculture) as well as to reduce other natural values and contribute to pollution. In Asia many rice fields were converted to shrimp ponds causing salination of adjacent fields.

3.9 *Heavy industry*

Industries are attracted to the coast where they can: (i) benefit from access to low-cost marine and inland transportation systems; (ii) gain the use of sea water for process or cooling purposes; (iii) find easy disposal of wastes into the sea; or (iv) obtain raw material from the sea. Heavy industry is also attracted by the labor sources and service conveniences of coastal population centers.

Coastal heavy industry engenders numerous infrastructural needs, such as dredging of ship channels which raises ecological problems. Also, coastal factories have often obliterated mangrove wetlands or tidal flats with seawalls, docks, landfills, and buildings. Wastewaters from heavy industries may cause damaging pollution from discharge of toxic substances including heavy metals (lead, mercury, cadmium, etc.), radioactive elements, acids, Polyaromatic hydrocarbons (PAHs), and innumerable other toxic industrial chemicals.

Heavy metals in the tropical marine environment originate with mining and dredging operations, factory operations, offshore oil drilling, desalination plant effluents, thermal power plants, and effluent and sewage discharge. A tragic example is the death of 1000 people at Minimata Bay in Japan (1950s) from eating seafood containing mercury which came from factory waste discharged into the Bay.

But much of industrial pollution can be eliminated by the application of existing, affordable waste treatment technology. Also, many estuaries, lagoons, and bays which are now polluted can be successfully reclaimed if the wastes entering them are adequately treated. Unified CZM can assist any existing pollution control authority though coordinated action.

3.10 *Ports and shipping*

Port infrastructure supports many activities, as per the following examples: (i) offshore oil and gas development requires port facilities, oil storage capacity, refineries, and other infrastructure support; (ii) the shipping industry requires channels, navigation aids, loading facilities, shipyards, and extensive land areas for container storage; (iii) fisheries development requires breakwaters, channels, ports, processing plants, and other facilities for middle and long range fishing fleets; and (iv) military operations require a variety of port facilities and support services.

Major threats from port development are: (i) preemption of fringing wetlands by landfill to create real estate for port expansion; (ii) dredging of channels and harbours (see Indonesia case, Section 8.9) and careless disposal of dredge spoils resulting in high turbidity, depressed oxygen, and sometimes the release of toxins (e.g. heavy metals); (iii) oil and chemical spills and persistent pollution; and (iv) induced urban stress with increased pressure on local infrastructure, institutions, and social systems.

It should not be surprising that pollution is at its worst in the harbors of large coastal cities with industrial ports. This pollution is particularly damaging to estuaries and lagoons because their sluggish circulation enables pollutants to reach high concentrations.

3.11 *Mining*

Mining for beach sand has been an important activity in many countries. Sand for construction is a valuable commodity but too many beaches have been depleted by mining. The key to the natural protection provided by the beachfront is the sand which is held in storage and yields to storm waves, thereby dissipating the force of their attack. Consequently, taking sand from any part of the beach or the nearshore can lead to serious erosion. For example, the beachfront receded 25 meters or more in one Caribbean event [14].

Alternatives to beach mining are available. Importing sand is recommended as the last resort because it means "exporting" the beach mining problem to another country and because the sand may contain pests, the sand type may not "fit" local needs, and importing is expensive. Offshore dredging of sand is usually not a good alternative because of the expense and ecological and geological problems but it needs to be explored. However, aggregate for concrete can be manufactured by crushing rock, gravel, or recycled glass or plastic to the appropriate dimensions, but financial incentives may be necessary. Replacement of fine sand for plaster is more difficult [14].

Coral mining requires similarly strong restrictions (Fig. 3). It was extensively undertaken in some island countries in the Indian Ocean and South-East Asia which diminished the natural storm barrier and left the shoreline exposed to erosion and storm surges, causing serious loss of beach and shoreland and damage to coastal human and marine resources habitats. Most countries are now prohibiting or seriously constraining coral mining (e.g. see Sri Lanka case, Section 8.15).

Mineral mining along the littoral can also disrupt coastal resource systems. Mineral mining (e.g. for tin) can be conducted from dredge barges

Fig. 3 Cutting coral blocks for construction, Bali, Indonesia. (Photo by J.R. Clark.)

in shallow water or from excavating equipment in sand-dunes and other lowland areas. Phosphate mining in shorelands and hinterlands produces phosphate wastewater which can cause considerable ecological damage when it reaches coastal waters.

3.12 *Petroleum*

Oil spills by tanker ships are a continuing threat to coastal seas ecosystems. However, few vulnerable countries have adequate contingency plans and emergency response procedures. Current international agreements have reduced oil pollution from shipping, but they still need stricter enforcement.

In one analysis of petroleum introduced into the oceans (off East Malaysia) 35% arose from marine transportation; 26% from river runoff; 10% each from natural seeps and atmosphere rain; 5% each from urban runoff, industrial and municipal waste; and 3.5% and 1.5%, respectively, from oil production and refineries [15].

Most countries have a national authority that attends to pollution, but CZM programs can offer valuable support in planning aspects.

3.13 *Shore protection*

People who build and live close to the water's edge face problems of serious erosion. When erosion is evidenced, the beach is often "armored"; that is, protective structures are built in an attempt to protect the threatened properties. But it has been recognized that: "Engineering structures, whatever their nature or type, have nearly always initiated, or aggravated, erosion in coastal areas." And further "... all structures and remedies, are temporary as only nature could halt coastal erosion" [16].

Shoreline constructions cause coastal zone disruption, particularly where hard structures are placed within the intertidal zone. Sea walls, bulkheads, groins, jetties, revetments, and breakwaters are built for the purpose of changing the shore regime which they do, sometimes quite dramatically, for better or worse. But these structures are expensive and often worsen the general erosion situation. Non-engineering solutions including setbacks and other planning approaches should be implemented.

While there are numerous examples of structures causing, or contributing to, destabilization of beaches, extensive areas of the coast are already occupied and must somehow be maintained safely.

3.14 *Tourism*

Tourism is one of the fastest growing sectors of international trade and a powerful tool for national development, particularly for smaller coastal countries and island countries with limited economic options. Tourism is said to be the world's largest single industry. It also has a large potential for social upset and for environmental damage to the Coastal Zone (see Solomon Islands case, Section 8.14).

Biodiversity reduction, resource depletion, and human health problems may result from the accumulated environmental effects of tourism. For tourism, the environment is itself a significant part of the product which a country has to offer. A successful tourism strategy will therefore seek to maximize the economic benefits of tourism development, while preserving the natural environment and the cultural milieu upon which it depends.

Beaches are the focal point for coastal tourism and a major source of hard currency for many countries. People travel thousands of miles and spend thousands of dollars to lie, sit, or walk on a beach. Good beaches are worth billions of tourist dollars. Degraded beaches discourage tourism. Unfortunately careless tourism development has ruined countless miles of beach, typically through building too close to the water's edge. When the sea threatens, bulkheads, groins, and seawalls are erected, which often make the situation worse (see Section 3.13).

A most serious impact from tourism development, worldwide, is decline of water quality. Sewage discharge is often the most common source of adverse effects on the biota and human health [17]. For example, in the Caribbean region, less than 10% of the sewage generated is treated and bacterial levels regularly exceed the international standard for recreational contact waters (200 MPN coliforms).

Oftentimes resource plundering and environmental deterioration for tourism are root causes of social disruption; for example, Dunkel [18] says that "... mass tourism subverts the indigenous culture" and that

". . . economic benefits have been inflated". If the situation deteriorates sufficiently, there is also loss of jobs, income, and hard currency earnings, factors that lead to socio-political instability. Such problems can be alleviated by appropriate planning mechanisms and development controls through unified CZM or similar approaches.

Many coastal countries are caught in the typical dilemma of tourism. They want the income from tourism while at the same time they deplore the negative social and environmental impacts. Clearly, in small countries dependent upon coastal tourism, disturbances do extend into the social and cultural scene.

Advocates of "ecotourism" hope that it can be used as a tool to deter the more rapacious resource uses—clear-cutting of forests and mangroves, — mining of coral reefs, filling of wetlands, using coastal waters as a waste bin—in favor of the lighter footprint of selective nature tourism.

3.15 *Water control and supply projects*

Usually the quantity and the natural cycle of water flowing to the sea should be maintained because freshwater admixture is important for sustaining many types of estuarine ecosystems. Many fisheries, such as shrimps, mullets, and oysters are strongly dependent upon river flows that enter estuaries and the sea. Rivers transport sand to beaches and beneficial nutrients to coastal biota (if not polluted). They establish optimal brackish conditions for mangrove forests and juvenile fish nurture places in estuaries as well as for nesting of colonial water birds.

Dams and water diversion or withdrawal schemes can seriously unbalance such river-dependent ecosystems and reduce their productivity and biodiversity by diverting water away from the coast or changing the balance of seasonal flows (the "hydroperiod"). For example, the famous Everglades of South Florida and the Sunderbans of Bangladesh have been seriously damaged by water control projects. Such interventions as store-and-release projects created for irrigation, flood control, or water supply, subject the coastal zone to surges of fresh water which typically cause disturbance of coastal ecosystems.

3.16 *Electric power plants*

Typical power plants with "open-cycle" cooling systems adversely impact coastal waters through thermal pollution which affects the aquatic ecosystem in the field of the outfall of cooling water discharge. Often an even greater threat is the high death rate of organisms suspended in the water and drawn into power plants with the cooling water.

Aquatic forms drawn in with the cooling water are exposed to heat, turbulence, abrasion, and shock. The potential for environmental damage from massive "entrainment" and death of these organisms—fish, plankton, and the larval stages of shellfish—is so great that it should be a major factor governing the design and location of power plants in the coastal zone. Where the water intake is screened, the number of fish impaled on the outside of these screens can be dramatic — five million fish killed in a few weeks in one example [6].

Most threatened are enclosed basins—estuaries, bays, lagoons, and tidal rivers — which are of critical environmental concern because of their high productivity and the abundance and diversity of life that they support, including nurture of the young of fishes and invertebrates. These areas, which require the highest degree of protection, are also the most attractive as power plant sites for many strategic reasons.

The major solutions are to: (i) locate power plants along the open coast, where there is deep water nearby for placement of intake and outlet structures if open-cycle cooling is to be used; and (ii) reduce the volume of cooling water used by plants on estuaries by requiring closed-cycle systems which reduce water intake by 95% by recirculating cooling waters.

Coastal power plants may pollute with any of the following toxicants: acids, acrolein, arsenic compounds, ammonia and amine compounds, boron, carbonates, chlorine and bromine, chlorinated phenols, chromates, cyanurates and cyanides, hydrazine compounds, hydroxides, metals and their salts, nitrates and nitrites, potassium compounds, phosphates, silicates, sulfates, sulfides, and fluorides [19].

3.17 *Nature's impacts*

The sea is not altered by man alone—Nature is implicated in many serious disturbances. Coastal managers should keep these effects in mind while combating human induced impacts. Examples of paramount concern include the following: "red tide", paralytic shellfish poisoning (PSP), starfish and ecological diseases which degrade coral reefs, sea level rise, natural subsidence of shorelines, and oceanic heating (El Niño). Add to this, cyclonic storms which sink ships, damage property, cause death, erode shorelines, and destroy coral reefs. Hurricane destruction to Caribbean coral reefs has been serious along with bioerosion.

Coral ecosystems have been disturbed by natural outbreaks of disease; for example a massive die-off of the long-spined sea urchin throughout the Caribbean in the early 1980s left the reefs without this natural defense against algal takeover (urchins eat algae which otherwise overgrow corals). Other outbreaks, such as blackband and whiteband disease and coral bleaching, are cause for increasing concern. Chemical disturbance of the ocean is associated with high mortalities of seabirds, such as brown pelicans (Monterey, California, USA, 1991).

One of the South Pacific's worst biological disasters is the widespread invasion of coral reef systems by *Acanthaster planci* (crown-of-thorns starfish) which have devastated large parts of the Great Barrier Reef (Australia) and other reef tracts.

The phenomenon known as "red tide" carries the threat of sickness (or even death) from "paralytic shellfish poisoning" (PSP) to those who are unfortunate enough to eat bivalve shellfish (mussels, oysters, clams, scallops) infested with dinoflagellate algae of the genus *Pyrodinium*. The risk is particularly high when the sea becomes red from the spread of these dinoflagellates. When eaten by bivalves, the dinoflagellate poisons are "stored" in the digestive gland of the shellfish which then can be eaten by humans.

Some red tides are toxic to fish and massive kills occur. Also, such marine mammals as bottlenose dolphin and manatee have suffered massive die-offs on the US Atlantic coast and dolphin in Mexico's Sea of Cortez, all attributed to a kind of red tide. In 1997, a new microbe (a dinoflagellate called *Pfiesteria*) was responsible for massive fish kills on the US Atlantic coast and even sickness of laboratory workers culturing the microbe.

Such problems are worldwide and in many cases associated with onshore development activities, particularly those involving excavation (yielding fugitive silt) or eutrophication (pollution by nutrients from sewage, detergents, etc.).

To add to these troubles, sea level has risen already, more than 30 centimeters in the past 100 years in some coastal areas. Sea level rise is associated with serious shoreline recession, cyclone exposure, and flooding along thousands of kilometers of shoreline. Further rise is expected under the influence of global warming, caused by release of carbon dioxide (and other gasses) into the atmosphere and the melting of glaciers, heating (and expansion) of the ocean water mass, and meltdown and disintegration of the Polar ice caps. Low-lying communities have to choose whether to retreat (e.g. move structures back) or entrench (e.g. build dikes).

4: Program Design

A tailor who makes a coat to fit a man is a useful person, but a tailor who should make a coat that would fit all men would be a genius. [Thomas G. Bowles, 1886]

Unified Coastal Zone Management (CZM) is a system for resource management operated by governments, but responsive to citizens. CZM combines development management (siting, structural codes, infrastructure routing, etc.) with resources management (resource conservation, environmental protection).

The primary CZM program utilizes the *regulatory powers* of government to achieve resource conservation, by controlling development activities and resolving potentially divisive conflicts among competing users of the coastal zone. Natural area conservation (the "custodial" approach) can be accomplished by exercising the custodial authority, or *proprietary rights* (rights by virtue of ownership) of government through declaration of protected areas, such as resource reserves, natural areas, and/or national parks.

The four most common and persuasive motives for CZM are said to be fisheries productivity, increased tourism revenues, sustained mangrove forestry, and security from natural hazard devastation [20]. Management may be initiated in response to a planning mandate but more often because of a crisis, such as a use conflict, a severe decline in a resource, or a devastating experience with natural hazards.

Unified CZM should be seen as a multi-sectoral process created to improve development planning and resource conservation though integration and cooperation. It should not be seen as a substitute for uni-sectoral programs such as tourism or maritime administration, nor as a substitute for coastal forestry or agriculture programs.

Fully unified CZM discourages piecemeal approaches to coastal development in favor of a balance between a variety of compatible uses whereby economic and social benefits are maximized and conservation and development become compatible goals. This chapter discusses the need for such a broad approach to coastal resources conservation and presents a methodology for the unified CZM process.

4.1 *Purposes*

CZM focuses on conservation of the *waters* of the coast but to do so it must control certain land based activities. Its basic purposes are to sustain coastal resources, conserve biodiversity, protect the littoral environment, and counter natural hazards. It accomplishes these goals mostly by influencing forms of shoreline development through education, resource management regulations, and environmental assessment but also by custodial protection of coastal resources (nature reserves). All these activities require coordinated community action for their accomplishment, a need that CZM fulfils.

Some of the most destructive "development" practises — massive coral reef mining or mangrove forest cutting — will usually have to be curtailed because these are not sustainable and they conflict strongly with other economic uses such as fishing and tourism. On the other hand, development activities with a lesser potential for damage can be adjusted in location, design or scale in order to meet multiple-use guidelines.

4.2 *Focal points*

In the process of formulating a unified CZM program, there are certain important focal points, as discussed below.

Issues. CZM programs are created to solve problems and to resolve issues which have been identified through a participatory process. Effective progress can only be made when the issues are clearly identified.

Objectives. Tangible objectives should be specified in the CZM Strategic Plan; for example, support of fisheries, protecting the community from storm ravages, attracting tourists, promoting public health, maintaining yields from mangrove forests, and preserving coral reefs. The objectives should be based on a set of issues the program can reasonably be expected to resolve within the time and within its available fiscal and staff resources.

Approaches. The main approaches of CZM are: (i) governmental regulations which protect biodiversity and control the harvest and use of natural resources; (ii) environmental assessments which can predict the impacts of various economic development schemes; (iii) establishment of resource reserves; and (iv) programs of resource restoration (see Section 6.8).

Multiple use. Many coastal uses could coexist in a multiple use approach while others might not, or would have to be severely restricted. The CZM role is to sort them out and recommend the optimal mix.

Participation. CZM programs require a high level of participation by "stakeholders" (those to be affected by coastal resource management). The general public of communities, along with private sector interests, are consulted about major coastal developments and the decision-making process is shared with them, requiring efficient communication and effective dialogue (see Philippines case, Section 8.13).

Coordination. The unification of CZM requires both integration of agency programs requiring intergovernmental coordination among agencies at all levels from central to community governments. Also essential is cross-sectoral coordination.

Planning. Without planning, CZM would be inefficient and any hope of integrating the various sectors and agencies would be lost. The essential planning element is Strategy Planning which precedes Master Plan creation.

Information. The key to success in unified CZM is appropriate compilation of information on resources, ecosystems, users, and impacts.

Incentives. For most countries, the motivation for implementing an CZM-type program will be very practical. Economic benefits will have to be shown. Explicit and persuasive social benefits will also be helpful. The values that developed nations put on biological diversity are not so fully embraced by some developing countries unless there are clear practical benefits, like attracting tourists.

4.3 *Program scope*

While at present there is no one correct model for unified management of coastal zones, there is available a family of coastal zone related management schemes that combine shorelands and coastal seas [1].

Structure. CZM programs require a minimum of six components: (i) legal, policy and administrative structure; (ii) project review and environmental assessment function; (iii) coordinative function; (iv) enforcement powers; (v) a community involvement component; and (vi) an educational and outreach activity. The program can start simply and elaborations can be added as appropriate, including: habitat restoration, technical assistance/fact finding unit, inspection, forward planning mechanism, extension services, research, protection of Special Habitats, survey and mapping, training, and so forth.

Work elements. Program activity includes policy making, issue identification, boundary definition, planning, zoning, land use control, data collection and monitoring, development guidelines and standards/codes, construction permits, environmental assessment, ecosystem restoration and nature protection, along with appropriate control measures, incentives, and disincentives. Specific planning processes are discussed in the following section.

Role. CZM is not intended to replace any existing governmental functions which are working well, but rather to strengthen them by providing the means for interagency and cross-sectoral coordination and by providing a unified strategy for coastal conservation in complex developments like tourism [17]. The CZM program is expected to focus most sharply on management of the physical development process using planning procedures and government regulations. Day-to-day resource management will usually have been mandated to other agencies even before the CZM program is implemented; for example, management of fisheries, mangrove forest harvests, marine parks and nature reserves, and pollution control. But CZM can be helpful; for example, it can help fishery managers by controlling adverse uses which impact fisheries stocks, such as discharge of pollution and destruction of the Special Habitats which serve as young fish nurturing areas.

Neither is ocean management typically a focus of CZM, although in some countries ocean management and CZM may be brought closely together (e.g. Brazil). Issues of greatest concern include shipping, offshore fisheries, mineral exploration and development, oceanic pollution control, and ocean research. International agreement on the "Law of the Sea" regime has stimulated initiatives in national ocean management.

4.4 *Program organization*

Unified CZM proceeds according to accepted modes and in a more or less standardized sequence. While the terms used may differ from program to program, the basic steps are about the same everywhere.

4.4.1 MODES

In its *planning mode*, unified CZM examines the consequences of various development actions and proposes necessary safeguards, constraints, countermeasures, and development alternatives that will guarantee sustainable use of coastal natural resources at the most productive levels possible. The two major types of planning for CZM are *strategy planning* and *master planning*. Also, some types of parallel planning are often involved.

In its *management mode*, unified CZM guides development in the coastal zone to promote resource conservation and biodiversity protection using a variety of approaches. CZM coordinates economic sectors to ensure that advances in one sector do not bring reverses in another; for example, that port development does not diminish local fisheries or tourism.

The concepts of CZM resemble those that underlie standard land use or economic planning and management approaches. So even though CZM-type programs and techniques are distinctive in theme and purpose, they are rooted in such well known and standard approaches as regional development planning, watershed management, rapid rural appraisal, and protected areas management.

4.4.2 SEQUENCE

While each country's program will be unique, there are several basic stages in the generation of a CZM program that are common to most, in one form or another. These stages are as follows.

1 *Policy formulation.* Creation of an initial policy framework to establish goals and to authorize and guide the CZM program accomplished by executive and/or legislative action, based on clearly defined issues. This stage includes authorization for strategy planning.

2 *Strategy planning.* Sometimes called "preliminary planning", this is the stage where the needs, feasibility, and potential benefits of the unified CZM policy action are explored and the basic strategy for the program is decided and articulated in a Strategy Plan. Information is the key.

3 *Program development.* In this stage, the CZM program is organized for implementation. A Master Plan for the coast is prepared, institutional mechanisms are created, and program responsibilities are assigned.

4 *Implementation.* Once the Master Plan and program details are approved and a budget and staff are authorized, the program can be implemented.

5 *Oversight.* Retrospective aspects may include formal evaluation and plan and program revisions.

In practice, the above stages are not so discrete and linear as theory suggests. Instead, there will be feedbacks and revisions of earlier stages as new facts and opportunities come to light in later stages. The whole program must be flexible and adaptable.

4.4.3 POLITICAL SUPPORT

The maximum of commitment to the planning *outcome* should be secured from decision-makers at the beginning, before and during the strategy planning phase. To be successful, the CZM program requires specific kinds of direction and support from governments. Particularly important needs are: (i) clearly articulated policies that match national development objectives; (ii) clear assignment of responsibilities and designation of a lead agency to formulate and coordinate the program; (iii) power to enforce rules and to achieve intergovernmental coordination; and (iv) funds to implement CZM.

Political and financial support is dependent on the level of awareness of decision makers. Political leaders should understand the cross-sectoral scope of coastal management and the consequences of resource mismanagement as well as the importance of the maritime dimension in national planning and development strategies.

The burden of proof for CZM advocates is to show strong benefits. In so doing, CZM programs should emphasize how to optimize the long-term economic benefits from the coastal zone and how to guide development and use of the coastal area in a rational and efficient manner for the highest benefit to the greatest number of people.

4.5 *Strategy planning*

Strategy planning is the key step in the process of organizing a unified CZM program because here the whole strategy is worked out. In this step the main "hows", "whys", "wheres", "whens" and "whos" are decided and methodologies chosen. As the initial planning step, strategy planning considers problems and opportunities regarding resources, economic development activities, and societal needs in the coastal area and devises a strategy to accomplish the CZM objectives. Of greatest importance, it sets priorities for action.

4.5.1 THE APPROACH

As the core of unified CZM, strategy planning explores options and develops an *optimum strategy* for a CZM program. It examines the facts, considers the issues, suggests possible solutions, and proposes specific legal and institutional arrangements according to the stated policies and objectives. Strategy planning involves all the preliminary investigation, data collection, issues analysis, dialogue, negotiation, and draft writing that is necessary to enable the government to define the problems, to understand the options, and to proceed to authorize a specific CZM program. The Strategy Plan lays the foundation for the future legislation or executive order needed to authorize the CZM program.

The strategy planning stage is where CZM program benefits are evaluated, where a wide array of data is accumulated, where a general strategy is created, and where recommendations are made for implementing policy and organization and administration of the CZM program. In strategy planning the potential impacts of the CZM policy action are explored; for example, on resources and resource users, on income and jobs, on social and cultural well being.

The Strategy Plan should: (i) address the issues; (ii) assign responsibility for the program to a particular agency; (iii) authorize the funding necessary for program development; (iv) state clearly the objectives of the CZM program; (v) recommend a method for collaboration among the various governmental agencies and private interests involved; (vi) state the time limits involved for various stages of program development; and (vii) require a specific program development, or tactical planning and organizing process.

The product, the Strategy Plan, is created in a way to answer questions in the minds of decision makers in government and leaders of stakeholder groups. The answers will lead to decisions to authorize or not to authorize the unified CZM program or to request more fact finding. Therefore, the information needed for the Strategy Plan is that which will enhance the decision-making process and that clearly depicts the trade-offs between short term sectoral gain and long-term national benefit.

Like any structural model or plan for governance, the approach described here is more instructive than prescriptive. Each country will have to choose and develop a unified CZM structure that is most suited to its own program needs and national political style. CZM should closely reflect broad governmental environment, social, and economic policy. For example, in countries with central planning, CZM initiatives should be included in each 5-year economic development plan.

4.5.2 BALANCE

The knack of strategic planning is to devise a program that will promote balance between economic development and the long-term environmental and socio-economic needs of the community. Compatible multiple-use objectives should always be the main focus. The maximum long-term flow of natural goods and services from a coastal resource system can be expected from a multiple-use CZM approach. The strategic plan should establish a method to eschew short-term development tactics in favor of long-term development and resource conservation strategies.

The idea is to accommodate as many uses as possible in the Coastal Zone using management controls to ensure compatibility and sustainability. Under CZM most coastal areas would be open for a variety of compatible uses. Only certain resource reserves, Special Habitats, and other specially designated areas would be set aside for restricted or single-purpose use.

4.5.3 UNIFICATION

The unified CZM program is designed as a package of planning, regulatory, and scientific tools along with economic development management. It is operated by governments (local and national) as a cooperative and coordinated venture in order to unify approaches to conservation. The CZM institution as an "overlay" agency, acts as a coordinating and consultative unit.

CZM is flexible. It can concentrate on the hazards of coastal erosion, as in Sri Lanka's CZM program (see Sri Lanka case, Section 8.15); on fisheries, as in the emerging Philippines program; on coastal and marine protected areas, as in the proposed Saudi Arabia CZM strategy; on shrimp aquaculture, as in Ecuador; on a "networking" approach, as in Oman (see Oman case, Section 8.12), or on land use, as in the United States (see USA case, Section 8.18).

To take an example, fisheries, tourism, oil and gas development, and public works may all be attempting to use the Coastal Zone simultaneously [20]. Both fisheries and tourism depend to a large extent on a high level of environmental quality, particularly coastal water quality. Both sectors may receive "spillover" impacts such as pollution, loss of wildlife habitat, and aesthetic degradation from uncontrolled oil and gas development. In another example, fisheries may require port services similar to those that tourism depends on and an infrastructure system that supplies water, sanitation, transportation, and telecommunications. Therefore, planning for both sectors should be integrated with that for transportation and public works sectors and consider the broader social and economic issues.

At the strategy planning stage it is most important to design a strong coordinating mechanism to ensure the widest and most effective participation of stakeholders—government agencies, the business sectors, communities, and the public at large (see Section 7.1). Local governments must be involved because that is where resources are found, and where the benefits or disbenefits are mainly to be felt. The central government has to be involved because responsibility and authority for marine affairs inevitably rests there (navigation, national security, migratory fish, international relations, etc.).

4.5.4 POLICY FRAMEWORK

In order to initiate policy action, it is necessary to make a persuasive case that serious coastal zone problems exist and that they can be solved. For example, CZM advocates should be able to answer such questions as the following.

1   Which coastal resources are seriously degraded; to what level have yields fallen; what are the economic consequences; and what actions are needed to correct this situation?

2   What are the causes of resource degradation; what type of developments and activities need to be controlled; what are the economic effects of the controls; and in consideration of the variety of possible trade-offs and their effects, what actions are recommended?

3   Who are the principal users of coastal renewable resources; how many jobs are at stake; how much income and foreign exchange earnings are involved in tourism, fisheries (Fig. 4), and other resource dependent industries; what further losses are expected if CZM is not implemented?

4   What are the *priority* issues; what Special Habitats need special protection; what sources of pollution are especially damaging; and what is the best approach—regulation or protected areas?

Policy statements should declare in the strongest terms possible that it is in the public interest to review and exercise control over activities that negatively affect the sustainability of coastal renewable resources [21]. In many countries a considerable amount of policy affecting coastal renewable resources will be in existence already. Therefore, an evaluation of existing policy should be a first step.

An example of *general* policy commitment for initiating a unified CZM-type program is the following statement formulated for the Philippines [22]:

Fig. 4 The fishing fleet at Lake Bardawil, Sinai, Egypt. (Photo courtesy of W.J.M. Verheugt.)

It is hereby declared a policy of the State to pursue a continuing program of effective management of coastal zones to meet the socio-economic development needs of our country for the benefit of present and future generations. In pursuing this policy, it shall be the responsibility of all government bureaus, agencies or instrumentalities, including political subdivisions involved in coastal management, to instill awareness to the public about the dangers of the degradation of environmental conditions in the country's coastal zones and encourage active participation of the people in all undertakings to conserve and enhance the country's coastal zones.

4.5.5 GOALS

The unified CZM program should be organized around a very specific set of goals and objectives generated in an open participatory process involving various levels of government and the stakeholders. The final declaration of goals must be made at the highest level of government.

The identification of broad goals is necessary to establish consensus on the direction the program should take, provide the framework for horizontal and vertical integration, coordination of coastal planning and management by the state/provincial, regional, and local governments and provide the criteria for program monitoring and evaluation. The set of goals forms the background and philosophical position against which more detailed policies, planning schemes, and management plans are prepared.

In a program of coastal management, the conservation issues and conflicts must first be identified before management strategies can be formed and specific countermeasures proposed. This identification process, known as "issues analysis", is the most critical aspect of strategy planning. Issues analysis is the anchor point for the CZM program, affecting goals and objectives, priority management needs, resource protection and restoration needs, identification of the stakeholders (who should participate in the program), information and research needs, the institutional arrangements needed, and uses to be permitted or prohibited.

Because CZM is an issue-driven process, the nature of the particular issues will dictate the type of program to be created. There should be a good fit between the set of issues which the unified program is attempting to resolve and the administrative mechanisms set up in response to these issues [20]. It is not enough to simply identify and list the issues. Each issue should be evaluated for at least the following: (i) the extent of socio-economic disturbance and resource loss that it causes; (ii) the degree to which it could be resolved by a CZM approach; and (iii) the consequences of not resolving it.

The management issues to be addressed in a CZM program can be so numerous that management may have to be prioritized in a kind of "triage" exercise. The easily resolved issues can be put in a high category because they can be worked out with minimum effort. The intractable issues can be given a low priority on the basis that much effort would be spent on them for the gain expected. Most effort would then go toward the realistically resolvable issues in between.

4.5.7 PROJECT REVIEW

The first step in the permit review process is prediction of impacts of proposed development projects utilizing Environmental Assessment (EA) which is linked to the development permit. The information and analyses generated in the EA provide the basis for the decision on whether a coastal project is disapproved, approved as is, or approved with conditions attached. In addition to environmental review, the natural hazard risk of the project and social and economic impacts are often considered (see Section 6.1).

Clearly, an effective project review system must be designed during strategy planning. A major purpose of CZM is to examine proposed major development projects via EA to determine the impacts they may have on coastal resource systems and to recommend design or location changes which can eliminate or reduce any negative impacts. The system is simple: if one wishes to build a structure, reclaim tidelands, clear land, or otherwise engage in a major coastal economic development activity, a permit must first be issued by government. How this shall be done should be made clear in the Strategy Plan.

4.5.8 BOUNDARIES

The Coastal Zone has both inland and oceanward boundaries. There is no single standard for placement of these boundaries. Where they are drawn depends totally on the issues the CZM program is supposed to address. But, all programs must straddle the shoreline, including some part of the sea and some part of the land. This transition area is where sea becomes land and land becomes sea, where government agency authority changes abruptly, where storms splash the land, where waterfront development locates, where boats make their landfalls, and where some of the richest aquatic habitat is found. It is also the place where terrestrial-type planning and management programs are at their weakest. What you can see while standing on the beach is the core of the coastal zone.

The Coastal Zone of some programs will be quite narrow because it is designed to address the littoral only. But a wider zone is usually better at the planning stage, one that includes a good piece of the land side of the Coastal Zone (above the high tide mark) as well as the water side. This may include shorelands that are pollution sources or where development could degrade Special Habitats or otherwise must be controlled to conserve coastal resources. To the extent possible, CZM planning boundaries should be functional boundaries, encompassing natural ecosystems and natural forces (floodlands, wetland ecosystems, etc.) but compromised to respect existing geographic and political boundaries [23].

For the inland boundary of the land side of the *planning* area, it might be convenient to use a major highway paralleling the coast, the foot of a coastal mountain range, the inland boundary lines of the coastal counties or municipalities, or other recognized political or physical feature. This might be more practical than an arbitrary distance of 100 meters or 1 kilometer and would be convenient for regional planning and social, demographic, and economic analysis as well as for defining stakeholders. The idea is to capture all the CZM issues within the identified coastal zone.

Pragmatically, the easiest approach to boundaries does seem to be drawing lines parallel to the shore at a set distance shoreward and seaward. As an example, the Sri Lanka CZM program has authority over a zone that extends from 300 meters landward of the tideline to 1 kilometer seaward of the mean low tideline; the landward boundary extends to a maximum of 2 kilometers inland where rivers, lagoons, or estuaries occur (see Sri Lanka case, Section 8.15).

On the water side of the Coastal Zone, the boundary should include the intertidal areas and nearshore waters, including all the coastal floodplains, mangroves, marshes, and tide-flats as well as beaches and dunes and fringing coral reefs. Most countries would want to extend the CZM boundary to the outer limit of artisanal fishing (30–40 meter depth), i.e. out to the maximum depth in which a small boat can anchor. Some countries might have reason to extend the boundary to the limit of the country's territorial limits (22 kilometers offshore). In any event, the concept of coastal zone should reflect all the different legal regimes within its waters, such as internal waters, territorial seas, contiguous zone, Exclusive Economic Zone, and also, where useful, straights and islands.

The narrower the coastal zone is defined, the more authority CZM can expect to gain. The broader it is, the less authority CZM can gain because of the appearance of duplication and competition with existing authorities, as well as a vagueness of function. Small island countries may have a particularly difficult time determining CZM boundaries. Some authorities would call entire islands coastal zones because most island commerce and societal affairs have a coastal connection (see Trinidad and Tobago case, Section 8.17). But calling a whole island a coastal regime and abandoning the concept of Coastal Zone as a definable and separable entity may jeopardize creation of a viable CZM program.

4.5.9 ZONES OF INFLUENCE

While it would be conceptually valid to include in CZM all areas that have an influence on coastal waters, this definition could be politically self-defeating if it attempts to encompass all coastal plains and the watersheds of all streams and rivers that drain into the sea (which at times extend hundreds of kilometers inland) as well as all the sovereign areas of sea. Political resistance to such a broad area of hinterlands and oceanic control might be insurmountable.

However, it is necessary to have some input into how resource uses are controlled in the watersheds which are the sources of excessive siltation and chemical pollution of coastal waters. This can be accomplished by the designation of a Zone of Influence (ZOI) lying outside the defined Coastal Zone but which needs management attention. The ZOI would be accompanied by a formal method of negotiating CZM conservation needs with a variety of hinterland interests. The ZOI approach has been used in the Gulf of Mannar (India) to formally negotiate management actions *outside* the boundaries of the official coastal reserve and which affect the reserve. In another approach, the Spanish Shores Act of 1988 identifies a ZOI of 500 meters from tidal waters [23].

The ZOI approach could work for offshore waters that extend past the statutory Coastal Zone boundary in order to cover the continental shelf (as defined by the 1982 Convention on the Law of the Sea) or the Exclusive Economic Zone to 320 kilometers offshore if important land use/water use issues extend that far seaward.

4.5.10 TIERS

In a subdivided approach to planning, the coastal zone area can be divided into "tiers", each representing a management unit characterized by different resources, issues, and jurisdictions. The purpose is to sharpen the unified CZM program by subdividing the Coastal Zone into planning subunits that conform to both ecological and jurisdictional boundaries. A four tier approach is shown below.

1 *Marine and coastal waters*. The deeper, open water part of the Coastal Zone seaward of the boundary of the transition area — high level of central government interest and authority, regional and international aspects.

2 *Transition area*. The core of the Coastal Zone (the littoral), defined to include all the nearshore and intertidal area of the coastal commons — high level of interest and authority by central, regional, and local government.

3 *Shorelands*. Shorefront lands directly adjacent to the transitional area which generate significant impacts to coastal resources, including storm impacted areas — valuable land, high level of local interest and authority.

4 *Hinterlands*. Uplands with critical linkages to the coast through soil erosion, pesticide or herbicide runoff, alterations of hydroperiod caused by dams and reservoirs, etc. — wide level of interest and authority.

Tiers 1 and 4 could be identified as ZOI, with Tiers 2 and 3 designated as the statutory Coastal Zone.

4.5.11 SETTING THE LIMITS

A unified CZM effort can be expected to grow gradually from a modest beginning to a more comprehensive program. CZM builds in small steps guided by a large vision (see Barbados case, Section 8.4). Each step is part of the larger plan and each step must succeed if the next is to start.

During strategy planning the individual steps, or increments, must be identified in a systematic manner. These can be organized according to functions (see Situation Management, Section 4.5.12).

The "function by function" incremental approach operates at the administrative level, whereby all the CZM functions are not implemented at one time, but rather are implemented function by function. For example, the program may start with project review and EA only; creation of reserves, habitat restoration, or providing technical services, could follow at later phases. The steps can also be organized by issues, or "situations", whereby individual problem situations are addressed one by one (see next section).

The incremental approach to CZM implementation — whether by function or situation — provides the opportunity to test concepts and approaches as a pilot effort before committing energies and political capital to a full-scale nationwide effort. The risks and consequences of failure are considerably less when a program is pursued incrementally.

An example of an incremental program is Sri Lanka. Here the CZM effort was initiated with a single issue then expanded to a more complete program. CZM started with a critical shore erosion situation and then expanded into a broader program; that is, large-scale coral mining enterprises in southern Sri Lanka in the 1960s and 1970s left the shoreline exposed to erosion and storm surges, causing serious loss of beach and shoreland and exposure of the coast to storm surges. A local fishery collapsed, mangroves, lagoons, and coconut groves were lost to shore erosion, and local wells became contaminated with salt water.

Sri Lanka reacted by enacting a unified CZM program in 1982, with first emphasis on controlling coral mining to protect the shoreline and a later (1990) broad emphasis on ecological concerns, particularly natural resources conservation (see Sri Lanka case, Section 8.15).

4.5.12 SITUATION MANAGEMENT

"Situation Management" describes a variety of CZM approaches to the management of *geo-specific* coastal problems, rather than country-wide problems. It is a CZM approach which handles the one-of-a-kind type of situation; for example, complex aquaculture problems in a particular coastal area, or a commercial port that has continuing expansion difficulties, or an urban center that is degrading coastal waters. The term "special area management" is in frequent use as one type of Situation Management involving particular areas. Another popular concept, "community based management" is compatible with the Situation Management approach.

In the Situation Management approach, the unified CZM framework can take up problems either: (i) resource by resource; (ii) issue by issue; or (iii) area by area. These approaches come under the umbrella of Situation Management because they are responding to individual geo-specific situations. Such targeted efforts can later evolve into more full-scale national CZM-type programs. In any event they must be organized with the assistance of national governments.

A point of concern might be an urban center surrounding a major bay, lagoon, or estuary, particularly when major port complexes are often the focus of the greatest intensity of coastal resource conflicts, and the greatest environmental degradation. Other examples might include lowland agriculture, coastal rural settlements, tourism build-up along beaches, or natural spectacles like flamingo or ibis migrations.

4.6 *Institutional mechanisms*

Most of the management strategies discussed in this report require direct action of governments. There is only modest CZM experience with the non-regulatory, incentive approach to conservation, say offering exclusive rights to resources to communities or offering market based incentives to the private sector such as allocations, user fees, and tax incentives. In the future, such approaches may be more common.

For now, governmental regulation is the main approach, which begs the question: "Who should be the regulators?" Also, "Where should the CZM authority be lodged within the government institutional structure?" When the first unified CZM programs were initiated some 25 years ago, the power was put in central government, usually in the resources or planning departments. Now things are quite different with most experts recommending a diminished central government role and a much stronger community role.

Traditional land-based or marine-based forms of administration and planning can be effective *if* suitably modified for the coast. CZM uses standard legal and administrative approaches but the institutional framework differs from the usual government "line agency" and sectorally oriented type of organization. The CZM institution is most often an "overlay" agency, a coordinating and consultative unit.

The variety of CZM institutional arrangements is sufficiently flexible to match whatever administrative system a country has, but institutional strengthening may be required.

CZM should be designed to fit neatly into the existing government structure and its programs and departmental responsibilities.

4.6.1 CENTRAL GOVERNMENT ROLE

CZM programs involve a strong central government role for many reasons, not the least of which is that central government typically has most jurisdiction over coastal and ocean waters. Also, it is the primary steward of the coastal Commons and has the governing tradition and the financial resources needed. Nevertheless, CZM programs can, and should, operate at a sub-national level with the cooperation of central government.

It would be appropriate to place the CZM office within an existing agency that already has appropriate regulatory powers, such as natural resources, fishery, or environmental ministries. Few countries will have an interagency, or interministerial, entity already in existence which is positioned to take on a unified CZM program. Therefore, a CZM lead agency will have to be designated and given an interagency mandate to accomplish the planning and operational functions.

The national CZM program can operate as an add-on to an existing agency or as a special independent office. Or it can be split so that the coordinating and planning functions are independent but the regulatory aspects are handled by one or more existing agencies (Fig. 5). It is necessary to establish an interagency coordinating committee to ensure the widest and most effective participation of numerous government agencies.

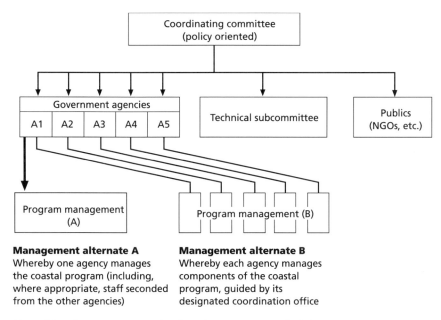

Management alternate A
Whereby one agency manages the coastal program (including, where appropriate, staff seconded from the other agencies)

Management alternate B
Whereby each agency manages components of the coastal program, guided by its designated coordination office

Fig. 5 Two alternative approaches for administering a unified CZM program. (Source: Clark [24].)

The CZM coordinating authority must have influence on the whole range of ministries and agencies involved, e.g. finance, agriculture, economic planning, commerce, tourism, forestry, and transportation. Clearly the greater the number of agencies and sectors involved, the greater is the potential for conflict. But, however it is done, there is no need to set up a new bureaucracy for a modest size program. However, for a full scale, comprehensive program, it might be desirable to create a new agency, such as a multiple-use oriented coastal authority (see Australia case, Section 8.2).

4.6.2 LOCAL GOVERNMENT ROLE

For CZM to work best, there must be decentralization, a sharing of political power with communities. The essence of decentralized CZM is that neither the coastal community nor the central government have exclusive power. There are many reasons central government must be involved in CZM (see Section 4.6.1). Provincial and local governments have to share authority, of course, because they operate closer to the people and local situations.

In a fully decentralized mode, central government has the lead role in matters of national significance and other centrally oriented matters. But otherwise it directs its attention to supporting local agencies that actually run the CZM program, often in a Situation Management framework. The popular jargon for this arrangement is "collaborative management" or "co-management" (see Section 7.4). Sri Lanka is a good example of a mature CZM program that has made the transition to decentralization (see Section 8.15).

4.6.3 LEGISLATION

The initial legislation or executive order needed to authorize the CZM program should specifically do the following: pronounce the goals and purposes of the CZM program; authorize funding; assign responsibility for strategy plan preparation to a particular agency; prescribe a method for collaboration among the various stakeholders (sectoral agencies, private sector, communities, and the affected public); and specify a step-by-step organizing and implementing process.

4.6.4 OPTIONS

Many countries may not, in the end, choose to go to an integrated, unified CZM approach, but rather adopt a less sweeping program with limited interventions. But, such countries are well advised, nevertheless, to reflect in their own approach the vision and goals of the unified CZM-type program.

   With the evolvement of CZM into a recognized field, it could become a common perception that countries without a formalized CZM program cannot or do not manage their coastal resources effectively. In some cases, however, it may be possible to achieve the same goals through an existing framework and infrastructure where government agencies responsible for different sectors of the Coastal Zone already have the potential to effectively coordinate their management activities. Examples are Singapore, the Maldives (where the entities are already well coordinated); and Trinidad and Tobago (where the central planning agency handles CZM matters). For further details see Maldives (Section 8.10) and Trinidad and Tobago cases (Section 8.17).

4.7 Strategy plan tactics

There are numerous matters to address in the strategy planning phase that will help to make the plan comprehensive. Some examples are presented below (not in any particular order).

4.7.1 COMMUNITY PARTICIPATION

It is an axiom of CZM that only a truly unified program (i.e. one that includes all the major sectors and interests affected) can accomplish all the needs. CZM can fail if important stakeholders are left out; for example, port authorities, housing departments, tourist industries, fishermen, tribal chiefs, economic development planners. Therefore, a major function of CZM is to provide a framework for coordination of a wide array of interests, most importantly the coastal communities and people that are directly affected (see Chapter 7).

4.7.2 LAND USE

Land use issues are usually considered in strategy planning and the outcomes included in the unified CZM Master Plan, perhaps as zoning schemes. This requires extensive coordination with a variety of stakeholders and agencies, including the physical planning office. The developed coastal zone faces intense conflict between private property-based operations in shorelands and public activities in the tidelands and coastal waters; for example, "reclamation" of mangrove areas for real estate or a factory placed at the shoreline. The unified CZM process has an important mediating role between "water side" interests and conflicting "land side" development interests whereby it seeks compatible solutions (see Section 6.11).

4.7.3 INFRASTRUCTURE PLANNING

Poorly planned roadways, bridges, airports, and other transportation infrastructure create special problems along the coast. These constructions often pollute the sea, preempt critical intertidal habitats, and obstruct natural water flows.

These infrastructure creations usually become development "corridors" or commercial "nuclei" which engender land clearing, filling, and construction, along with soil erosion and pollution. These effects may jeopardize coastal ecosystems and should therefore be considered in the knowledge that publicly supplied infrastructure controls the extent and direction of economic development.

Because of the importance of water circulation and flushing in lagoons and estuaries, activities that alter basin configuration or dynamics can create disturbances with far reaching effects. Major adverse effects stem from construction of causeways and bridges across bays and from dredging undertaken to create navigation channels, turning basins, harbors, and marinas. Other problems arise from laying pipeline or excavating material for landfill or construction.

Everything about infrastructure development should be carefully planned in accordance with conservation guidelines. Fortunately, planners can predict the type and extent of development that will occur along infrastructure corridors, roadway routings, terminal locations, and so forth, and can guide their location and design for the purpose of resource conservation. The CZM agency should prepare a set of guidelines to inform planners and engineers.

Development planners, must recognize that modification of the hinterlands has a high potential for adverse effects on resources of coastal zones, particularly in littoral basins. Examples of major impacts are: land clearing and grading, agricultural and forest clearing, runoff pollution (Fig. 6); siltation

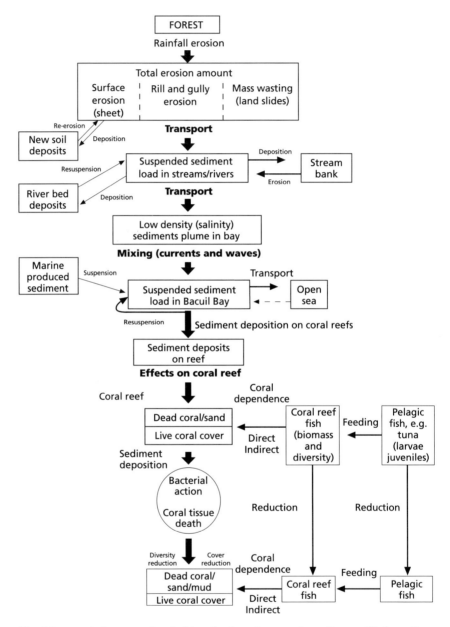

Fig. 6 Impact chain connecting the hinterlands to the coastal sea. (Source: Hodgson & Dixon [25].)

from eroded uplands; dams across major rivers; land filling to provide real estate for industry, housing, recreation, airports, and farmland; harbor improvement; and mining/quarrying.

While it has not been common for those who plan dams or land clearing enterprises to consult with coastal conservation interests, coastal resources are jeopardized by land use activities—when the terrain of the coastal watershed or its drainage is significantly altered. The suggested approach is to identify, as a ZOI, an appropriate area of the hinterlands that lies past the statutory inland boundary of the Coastal Zone and set up a formal coordinating mechanism (see Section 4.5.9).

4.7.5 POLLUTION

The Strategy Plan should recommend the level of involvement that the CZM program would have with pollution control. Aside from outright fish kills and other dramatic effects, such as human disease, pollution causes pervasive degradation that is evidenced by depletion of fish and shellfish, decline in biological carrying capacity, and uglification of tourist areas. Coastal seas are susceptible to pollution originating on land and carried to the coast by streams, drainages, pipes, sluices, and stormwater systems.

In most countries, a pollution control authority will already have been established and operating and would be expected to retain continuing responsibility for pollution in regard to ecological and human health. CZM should coordinate with that authority regarding conformance of development initiatives with national pollution standards and other matters. In this role, projects with unacceptable potentials for pollution would be discouraged or modified to a level of acceptability and conformance with standards.

Of particular concern to CZM are such diffuse sources of pollution as runoff from farms, forest clearing, and city streets. CZM might also be helpful in guiding the use of phosphate based detergents in coastal areas where the standards are (by weight): laundry detergents, not more than 6% P; dish-washing detergents, not more than 0.5% P.

4.7.6 HABITAT PROTECTION

CZM addresses conservation of natural habitat types known to be especially valuable; for example, mangrove forests, coral reefs, submerged seagrass meadows, kelp beds, oyster bars, beach-dune systems, and lagoons, estuaries, and other embayments. While it is useful and practical to focus on individual habitat types, one must not forget they exist only as components of wider coastal ecosystems. Custodial protection of Special Habitats is integrated into CZM programs so that managing a nature reserve or marine park does not have to be done in isolation from surrounding sources of pol-

43

lution or troublesome land uses, nor without interagency collaboration (see Section 7.2).

4.7.7 HABITAT RESTORATION

The CZM program should include a component focused on rehabilitation of Special Habitats that have been disabled such as mangrove forests, coral reefs, and lagoons. Degraded habitats should be rehabilitated to the highest practicable level of productivity and biodiversity. CZM staff should investigate the condition of coastal habitats and make an economic evaluation of benefits of restoration. Habitats to be restored should be identified, mapped, and listed with priority indexing.

4.7.8 SPECIES PROTECTION

A distinction of the sea is its limited endemism; marine species and subspecies are only rarely confined to certain small areas. Since few mobile species are confined to narrowly bounded habitats, the chance that any one might be extinguished by human activities is low, with exceptions [26]. Yet care must be taken to protect such species as are endangered and to conserve the genetic resource base.

Those species that are recognized as endangered are mostly big animals, including dugong, river dolphin, some whales and other cetaceans, seals and sea lions, along with crocodilians and several species of sea turtles. Also, there are a number of littoral birds recognized as endangered, such as terns, egrets, eagles, and pelicans. In 1996, a group of experts declared that more than 100 species of coastal/marine fish were "threatened" and that 15 species were "critically endangered", mostly by excessive commercial exploitation. The list included white shark, Nassau grouper, bluefin tuna, some sea horses, and most sturgeons [27].

Species protection is one of the major successes of the world environmental movement. The list of endangered sea and shore species that have been revitalized (brought back from low levels) includes: pelican, osprey, crocodile, alligator, gray and humpback whale, eagle, manatee, and many others.

4.7.9 NATURAL HAZARDS PROTECTION

The planning program should address natural hazards prevention and mitigation, including defenses against any sea level rise caused by global warming. In the Strategic Plan, recommendations should be made about inclusion of natural hazards prevention in the CZM program. This is critically important if no other governmental agency is dealing with the subject of maintaining storm defenses.

A CZM program is a good vehicle for combining coastal natural hazards

Fig. 7 In Bangladesh, cyclone shelters each give refuge to a thousand or more people during cyclone events; in the interim they may be used for schools. (Photo by J.R. Clark.)

(e.g. hurricanes or cyclones, typhoons) prevention with resources conservation in the coastal seas (Fig. 7). As many experienced planners and managers already know, the measures best suited to conserving ecological resources are often the same as those needed to preserve the natural landforms that serve as barriers to storms and flooding. Some common elements of resource conservation and hazard mitigation are as follows [9].

1 Both require integrated approaches and centralization of authority in order to control the location and type of development.

2 Both require preservation of the natural elements that protect coastal populations from cyclonic winds and storm surges, e.g. mangrove forests, sand-dunes, beaches, and coral reefs.

3 Both require management of coastal watersheds and river basins, e.g. in order to protect municipal water supply and coastal water quality and to reduce coastal flooding damage.

4 Both require involvement of many levels of government from national to local, as well as international cooperation in some instances.

Because officials responsible for coastal hazards concern themselves mostly with emergency response and post-disaster relief, the condition of the coast's protective resources has been too often ignored. That is why it is essential for CZM to undertake a primary role in protection.

4.7.10 BEACH PROTECTION

Troublesome erosion of beaches occurs in developed areas where buildings and roadways have been placed too close to the water's edge and are being undermined or threatened by storm induced erosion. Erosion can be countered best by keeping structures back a safe distance from the shoreline using "setbacks" (see Section 6.6). Thus erosion management first off becomes a land use matter. CZM recommends non-structural approaches like setbacks instead of massive seawalls and groins which are expensive and often self-defeating (see Section 3.13).

Since the main threat to the beach is usually from development on land next to it, beach protection requires coordinated management of the beach itself and the land behind it, as well as a way to limit buildings, prevent sand mining, and control beach protection and inlet structures. These needs can be fulfilled by a unified CZM program.

New "soft engineering" technologies are available, as an alternative to structural protection against storm waves, storm surges, coastal flooding, and sea-level rise. These technologies require less concrete and rock and are protective of coastal environments [28]. The solution is not to go exclusively with either structural or non-structural techniques but to achieve a balanced plan emphasizing the non-structural.

The storage capacity of each component—dunes, beach berm, forebeach, and beach bars—must be maintained at the highest possible level (see Section 2.6). Therefore, beach conservation should start with the premise that any removal of sand is adverse, whether for land fill, concrete aggregate, plaster, or any other purpose, and should be prohibited. The mining of beach sand and/or aggregate has been identified as a major problem in island countries; for example, most Caribbean Islands have lost important tourist beaches and are now prohibiting most sand mining (see Sections 3.11, 8.1 & 8.4).

4.7.11 LOCAL ASSISTANCE

An important aspect of unified CZM is the level of assistance that central government will provide to communities involved with the program. Provision of technical and financial assistance to communities to help them accomplish their CZM goals is often the best way to build enthusiasm and gain cooperation. For example, one CZM program in the United States (Florida) operates mainly through the giving of grants to coastal communities.

4.7.12 ALTERNATE EMPLOYMENT

In many cases CZM programs are being developed in which resources are restricted or reallocated, to the detriment of one or more stakeholder groups, such as artisanal fishermen who would be displaced by a tourist development, coral miners who would lose jobs if mining were controlled, or charcoal makers by a ban on cutting of mangroves. Coastal communities often depend on natural resources of the Coastal Zone for their everyday livelihoods. Such marginalized people do not have the financial cushion to risk management actions that might deny their means to a livelihood even for a short period. Some affected persons find other jobs, some find welfare, and others just suffer. CZM planning should address this situation. If such arrangements are not made, administrators may not be willing to enforce conservation rules because of negative social impacts.

A promising approach is the development and provision for alternative livelihoods guided by the CZM program. For example, in some places where fishermen or coral miners have been displaced, they are retrained to become tour guides (Bali, Indonesia; Gulf of Fonseca, Honduras) or farmers (Sri Lanka). This is not to imply that the CZM agency should become an employment agency. Rather, the institutional arrangement for CZM should include explicit linkage with the relevant agencies, such as the Ministry of Social Welfare or the Ministry of Economic Affairs, who already have responsibility for retraining and employment.

4.7.13 NATIONAL SECURITY

National security interests in the Coastal Zone are high because the coastal zone is the frontier, or border zone, and is under threat of invasion or other negative activities (e.g. smuggling). Naval ports and harbors, coastal airfields, and special bases of all kinds are sited in the Coastal Zone, usually with a high priority and unusual security and sometimes a high potential for resource damage. However, with the correct approach, the military can be expected to cooperate in coastal zone conservation so long as it is not in conflict with national security needs. Therefore, the defense establishment should be included as a party to any CZM program.

4.7.14 INFORMATION

Data collection and assimilation is an important part of the strategy planning phase. It is during this phase that the most important decisions will be made about the future of the unified CZM program, or even whether it will have a future. Clearly, these decisions should be made in the most data rich circumstances so that the consequences of taking or not taking specific actions are rational.

In the strategy planning stage, the information needed is often for convincing decision makers and stakeholders of the need to authorize a CZM program, or not to, if it is neither feasible nor appropriate. Activity will mostly be aimed at collection of secondary (existing) data, but allowance should be made for specific original data collection for orientation of the Strategy Plan (see Section 6.10).

Because the information needs for a CZM program depend upon the issues to be addressed and because these vary considerably from country to

country, it is not possible to set forth a standard list of information requirements. However, we have listed below information items that planners and
managers may need, particularly for strategy planning [1,24,29].

1   *Physical environment.* Terrain data, erosional processes, storm surge,
winds, tides, air–sea interaction, sediment transport, subsidence, sediment
supply, and condition of above.

2   *Biological environment.* Distribution and extent of living resources;
trends in abundance; major habitats, ecosystems, ecological relationships,
food chain; endangered species, indicator species, diversity.

3   *Coastal renewable resources.* Fisheries and aquaculture activity and
yields; mangrove exploitation; special habitats, e.g. mangroves and other
wetlands, beaches, dune fields, seagrass meadows, coral reefs, tide-flats, estuaries, lagoons, shellfish beds, and special nurture areas (breeding, feeding,
security); restoration needs.

4   *Users of resources.* Resource use patterns (including economic output,
jobs, revenue, investment and tax yields) including fisheries (activity and
yields), tourism, ports, shipping, energy, settlements, transportation, aquaculture, mining, oil/gas, waste treatment and disposal, traditional practices.

5   *Impacts.* Impairment of coastal resources; pollution, habitat losses,
species depletion, sedimentation, visual degradation, pollution; special
problem situations (e.g. highly polluted estuaries), mangrove clearing,
destruction of coral reefs.

6   *Upland effects.* Impairment of coastal resources from river dams and
diversions, accelerated sediment transport, reduction of freshwater inflow,
disruption of natural hydroperiod, reduction of beach nourishment with
erosion and pollution.

7   *Natural hazards.* Identification of areas of high risk from natural
hazards such as badly eroding beaches, floodable lowlands, landslide prone
slopes, etc.

8   *Socio-economic.* Economic statistics for coastal communities; social
organization; dependencies on coastal resources; historic use patterns;
factors determining uses and sustainability; demography; socio-cultural
data; social equity; land use.

9   *Institutional.* National, State, and local authorities; ministries and
departments with coastal interest; organization; non-government organizations (NGOs); legislation on zoning, pollution, resource use; interagency
councils; advisory panels; standing agreements with private parties; permitting and other administrative processes; capacities.

10   *Critical species.* Identification of coastal species of particular
significance or economic value or that are threatened with extinction and
their habitats; and remediation needs.

11   *Special habitats.* Habitats of critical importance such as mangroves and
other wetlands, beaches, dune fields, seagrass meadows, coral reefs, tide-
flats, estuaries, lagoons, shellfish beds, and special breeding and feeding

areas for coastal species; areas that should be designated as reserves or other protected areas; restoration needs.

4.7.15 FUNDING

Funding sources for the CZM program need to be identified early on. CZM programs in developing countries are eligible for initial financial support from international funding agencies, various international banks, foreign aid agencies, and other donors, particularly for the planning phase. Sustained financing for the management phase will likely come from internal sources.

5: Program Development

*Disregarding for the moment the question of moral
purpose, it is safe to say that the property of our people
depends directly on the energy and intelligence with which
our natural resources are used.* [Theodore Roosevelt, 1908]

The strength of the operational Coastal Zone Management (CZM) program
comes from its Master Plan which evolves from the Strategy Plan and is the
operative basis for the program. Its components reflect the need for the
program, the method and legislative basis, rule making, implementation
strategy, staffing and financing, public participation, Coastal Zone boundaries, permits and environmental assessment, and other relevant aspects.

In developing the CZM program from the Strategy Plan, initial emphasis
should be put on preparation of written guidelines for development management, environmental protection, and multiple-use planning.

5.1 *The approach*

The management operations phase of CZM can facilitate any variety of
integrated coastal management goals that address *joint* management of the
water side and land side of the coastal zone. There is no "standard" Master
Plan; its format is altered to accommodate the institutional and customary
systems of the countries involved, including political and administrative
structures, economic conditions, cultural/religious patterns, social traditions, and community needs.

The CZM management office should be mandated, staffed, and budgeted
to accomplish at least the following three tasks: (i) interagency and inter-
sectoral coordination on coastal resource conservation matters; (ii) review
and comment on all major coastal developments; and (iii) enforcement
of CZM rules and decisions. Its role is to make judgements about development and resources management that are for the good of the nation as a
whole and not for the good of a particular economic sector or agency of
government.

5.2 *Master Plan*

The Strategy Plan anticipates the Master Plan, designs its structure and
specifies a certain type of program to support it.

The Master Plan identifies options for economic progress in the coastal
area and identifies governmental and private actions to accomplish beneficial
and sustainable change in resource use; i.e. change that is economically
sound and socially just and that sustains the natural resource base.

The Master Plan identifies the permissible type of uses and the standards for these uses, the procedure for permit approval processes, for monitoring of activities and for compliance. It indicates options for human progress in the coastal area. For example, it recommends governmental and private actions to accomplish beneficial and sustainable change; i.e. change that is economically sound and socially just and that maintains the natural resource base. It should have a specific set of objectives as its foundation.

Substantive items for the Master Plan could include any of the following objectives:
1   Maintain a high quality coastal environment.
2   Identify and protect valuable species.
3   Identify and protect Special Habitats.
4   Identify lands particularly suitable for development.
5   Resolve conflicts among incompatible activities.
6   Control activities with adverse impact on Special Habitats.
7   Control pollution from point discharges and from land runoff as well as accidental spills of pollutants.
8   Restore damaged ecosystems.
9   Promote sustainable use of coastal resources.
10   Balance economic and environmental coastal zone pressures.
11   Recommend ecologically safe options for coastal development.
12   Raise public awareness about conservation.

Process and information items appropriate for the Master Plan might include any of the following (many of these arise from the Strategy Plan):
1   Existing government interventions.
2   Management mechanisms, e.g. permits, Environmental Assessment (EA).
3   Coastal Zone boundaries.
4   Major issues.
5   Authorities: power sharing for an integrated strategy.
6   Guidelines and standards for coastal development.
7   Goals, objectives.
8   Resources inventory (Fig. 8).
9   Land use suitability and project compatibility.
10   Indicated solutions.
11   Institutional arrangements: empowerment, linkages.
12   Wherewithal, budgets, external support.
13   Staffing, training, motivation.
14   International aspects.
15   Public participation.
16   Information needs.

The Master Plan has a fixed life expectancy because time changes all things and each version of the Master Plan is an interim document. Since the conditions will be constantly changing, the plan will be in need of periodic updating.

Fig. 8 Surveying a coral patch reef. (Photo courtesy of W. Keogh.)

5.3 *Project review and assessment*

At the center of the CZM management phase is a project review and permit system to ensure that major developments are compatible with conservation goals. The regulatory component of CZM that controls project review provides the broad administrative framework for conserving coastal resources, including rules, permits, performance standards, guidelines, etc. This component operates through the administrative process and the "police power" of the country and its provinces. Under the permit system, no development can begin until a permit is issued by the CZM agency. This regulatory function will require an EA type of fact finding procedure for each coastal project (see Section 6.1).

5.4 *Service function*

The responsibilities of the CZM agency includes production of considerable technical information, including surveys and delineation of special areas. Much of the output might be maps and overlays. Any of the following services might be provided by a full scale CZM program.

1   *Technical services.* Planning, information management, survey and mapping, monitoring, research, economic analysis, and technical advice to developers, government agencies, and interest groups.

2   *Education services.* Information dissemination, public education, staff training, extension services (see Section 7.3).

3   *Guidelines.* Formulation of standards and guidelines for coastal development, and publication and update of the standards and guidelines.

4   *Restoration advice.* Assessing the state of coastal renewable resources, determining restoration needs, designing restoration projects.

5 *Protected areas*. Determining the need for the establishment of a system of parks, reserves, or other types of protected areas, priorities for site selection.

The users of this information would-be developers and their consultants, design engineers, various government agencies and their planners, environmental interests, and the media.

With a CZM management unit reviewing development proposals and often requesting changes in project specifications, developers will soon find that they can benefit from prior consultation with CZM staff. Also, a need for technical descriptions and guidelines will emerge.

As noted, planning must be an ongoing activity. It is undertaken by the CZM technical staff with assistance from Universities, consultants, or research institutes (see Trinidad and Tobago case, Section 8.17).

5.5 *Environmental Management Plan*

An important asset for project-by-project management is the Environmental Management Plan (EMP). EMP was not created to slow economic development but to encourage it to be sustainable. The idea is to ensure that development and conservation can coexist [1]. The EMP responds to the sources of impact of a project as far as they can be known in advance and instructs the proponent regarding environmental matters. It addresses environmental management needs for the immediate site, for adjacent areas, and for areas offsite which are subject to project generated impacts, e.g. staging yards and borrow areas. The EMP also addresses secondary (linked) impacts, such as sewage or roadway traffic that would be induced by more business. Socio-economic impacts are also considered.

5.6 *Constraints and complexities*

The pathway to unified CZM is strewn with obstacles. All the usual resistance to government intervention may be there along with high levels of interagency strife and private sector interference and, oftentimes, low levels of scientific information and public support. Other implementation problems commonly include: lack of political will and support, inappropriate or inadequate administrative arrangements, failure to enforce the law and penalties, and penalties which do not act as disincentives [4].

Economics. In convincing supervisors, decision makers, and legislators of the need to create a CZM program, nothing is more important than persuasive economic evidence. National income, foreign exchange earnings, employment and local self-sufficiency are most important factors. Because of the importance of economic justification of CZM, it may be desirable to employ a professional resource economist to assist with economic analyses.

53

Contrary to some current impressions, conservation and economic development are not conflicting ideas. In fact, well planned, conservation-oriented development will add to the general economic and social prosperity of a coastal community, while bad development will sooner or later have a negative effect.

Patience. Trying to do too much too fast with a CZM program may be a mistake. Advocates of CZM should not overload the program, should avoid unrealistic goals and excessive complexity, and should be cautious and willing to negotiate. If the program appears too complex, too controversial, too disrupting, or too expensive, the government may stall the program in the planning stage. The realities of political, economic and social life of many countries may start with a less ambitious beginning than full-scale CZM but one which still strives for unified CZM ideals. Although building a CZM program is a long-term process, some tangible outputs and outcomes (benefits) must be shown in the short term to win over opposition and build a supportive constituency [4].

Opposition. In the initiation of a CZM program, strong, sectorally oriented agencies may be disturbed about their potential loss of power and autonomy that might result from creation of new institutions or reorganization of existing ones (see Section 8.4). Integration of multiple agency interests into a single program is difficult. Such public agencies as transportation, housing, military, agriculture, and industry may see the CZM program as a problem rather than as an opportunity. Also, CZM may be unpopular with certain profit oriented, powerful private sector interests such as manufacturing and aquaculture who initially may object to any controls — they should be assured that the CZM program will treat them fairly and that its functions are beneficial.

Training. Special training is needed in the skills required for CZM, particularly for fact-finding and managerial personnel who formulate policy and organize program elements involving skills in law, economics, and development management. Staff training is essential in a field like CZM where so many different types of activities and technologies are involved and where new findings and new technologies occur so rapidly; for example, in electronic data gathering, management, and analysis and in electronic communication. Because CZM works to resolve inter-sectoral conflict and achieve balance of uses, staff also need special training in CZM's cross-sector approach. Training might range from short courses and on-the-job training to formal university degree courses (see Section 9.11). International assistance is available to most developing countries to meet a variety of training needs.

Training programs should be developed on various aspects of coastal planning at national and regional levels, including: seminars and workshops

for policy-makers and planners; short-term speciality courses; development
of new academic programs aiming at the development of multi-disciplinary
competencies (e.g. law of the sea, environmental economy, socio-economy,
sociology, demography); and preparation of training materials (manuals,
simulations, audio visual aids) for schools and various strata of the public,
including the rural communities [29]. International assistance is available to
most developing countries to meet a variety of training needs.

In one example, the Coastal Area Planning and Management Division of
Trinidad and Tobago (located in the Institute of Marine Affairs) listed its priorities for staff training in CZM as: (i) assessment of sociological impacts; (ii)
assessment of economic impacts; (iii) cost–benefit analysis; (iv) simulation
analysis; and (v) resource management [24].

The foreign study tour is often a good approach. But such study tours are
often expensive and lose their value if poorly organized, or too general in
focus. A temporary assignment of 1–6 months in a particular planning office,
research group, university faculty or actively managed coral reef park, can
provide effective "on-the-job" training for mid-level staff. Language problems can be significant; English will generally be the necessary language.
Study tours are often inefficient if arranged for a single trainee who requires
substantial personal attention from host staff [30].

Corruption. It is very important to prevent graft and corruption in a CZM
system because certain persons may try to control the permit process for
their own enrichment. Bribes, payoffs, and other forms of corruption are not
uncommon in government regulatory programs and must be anticipated to
be avoided. To take an example approach, in 1997 Germany set criteria for
control of its foreign aid donor program in regard to corruption and civil
rights and how the recipient governments should act. Germany wants to
bring the corruption issue into the open and enter a partnership with the
recipient country to control it. On civil rights Germany asks for: (i) observa-
tion of political and civil rights; (ii) participation of the people in the political
process; (iii) guarantee of the rule of law; (iv) application of market
economy; and (v) government oriented to development. This "model" was
then accepted by the European Union and by Switzerland. Such moral oversight is commendable.

5.7 *Success factors*

The US Agency for International Development (USAID) supported a test
program for CZM involving countries in Asia and Latin America for more
than 10 years. The following are the strategies which, according to Olsen
[31], have led to success in CZM in these countries.

1   Work with all levels of government (central and local).
2   Focus on a limited set of issues.

3   Emphasize public education early in the program.
4   Gain strong participation of "resource user groups".
5   Utilize scientific information and monitor the program.
6   Emphasize training of coastal managers (present and future).
7   Move quickly from planning to management with sub-projects.
8   Encourage flexibility and adaptability.
9   Think of CZM as an experiment with a "steep learning curve".
10 Use an incremental approach (implement program in steps).

6: Methods and Tools

Give us the tools and we will finish the job. [Sir Winston Churchill, 1941]

Many of the methods and tools used in planning and administrative aspects of unified Coastal Zone Management (CZM) are modifications of those common to other pursuits while others are quite specialized. Some of the more important ones are discussed in this chapter.

6.1 *Project review*

An important aspect of CZM programs is project review preliminary to permitting action on a project proposal. Project review should incorporate relevant environmental, socio-cultural, and economic assessments.

6.1.1 ENVIRONMENTAL ASSESSMENT

Environmental Assessment (EA) has been used extensively during the last 20 years to assess the effects of coastal developments, usually as part of the CZM project review process. As an analytic method, EA is used to predict the environmental effects of a project or a program. As a government process, environmental assessment is imposed on both public agencies and private developers to predict environmental impacts and to coordinate aspects of management.

It seems indisputable that there must be a systematic way to review coastal projects so as to modify them to reduce negative impacts on resources and to reject the worst of them. According to Sorensen and McCreary [20] three fundamental benefits arise from these objectives: (i) cause and effect relationships can be determined with reasonable accuracy and presented in terms understandable by policy makers; (ii) prediction of impacts will improve planning and decision making; and (iii) the government can enforce decisions emanating from the environmental process.

The term Environmental Assessment is used here for the general process of assessment, and the term Environmental Impact Assessment (EIA) is used for the full scale analytic phase justified when the impacts of a project appear to be potentially severe. The EA used for coastal zone development projects is generally the same as the standard terrestrial or marine EA, about which there is an extensive literature. Of particular value to CZM are recent reports from UNEP Regional Seas [32] and the East–West Center [33].

The proponent of the project will be granted development permission if the impacts identified by EA are eliminated or reduced to a minor level or if they are appropriately "mitigated" (mitigation refers to actions taken to offset negative environmental impacts). Over time, experience gained with different kinds of projects and different resources will raise the level of confidence of the CZM program and actually enable it to speed up development by providing useful strategic guidance to coastal communities and to development interests.

It is usually best to have two levels of assessment; the first level would be a preliminary EA review to see if there are potential serious impacts. If so, the project would be subject to higher level, more intensive EIA, as is done in the "AMDAL" EA procedure operating throughout Indonesia.

The author has created and field tested a seven-step procedure called "STEPS" (Standard Track for Environmental Predictions) which works well for EA of coastal projects.

Step 1: Project identification. This is a simple description of the subject development project including textual (verbal) and descriptive (spatial, maps and plans) components. It should be focused on the aspects that are most likely to cause environmental disturbance. Later in the sequence it may be necessary to obtain more detailed descriptions of the engineering plans. If there are alternative approaches to the project, as there should be, each one is described.

Step 2: Resources at risk. This is a description of the particular resources at risk to the proposed project, including water, soil, plant, energy, and animal components. Analysts should not waste effort on trivial matters or resources *not* to be significantly affected by the project. For example, discuss climate *if* relevant, soils *if* involved, hydrology *if* implicated.

Step 3: Screening. This is the reconnaissance step where the analyst should identify all environmental problems that could be caused by the project and sort them out into two categories, major and minor. The minor ones are discarded with a brief explanation for each, and the major ones are passed on to the scoping step. "Checklists" and "matrices" may be used at this stage.

Step 4: Scoping. In this step you, the Assessor, make predictions of the major impacts and make a preliminary determination of levels of significance. You discover feasible engineering alternatives which result in lesser environmental impact and explore mitigation requirements. These initial findings are presented at special meetings, or by other means, to the stakeholders for comment. Practical alternatives to the project are identified. The results of these consultations along with backup technical information and any recommendations from the Assessor are passed to Step 5 *if* it is decided that a full scale EIA is needed.

Step 5: Analysis. In conducting an EIA, the assessor is guided by the results of the Scoping step. Additional data collections, analyses, and reviews are completed. Final estimates are made of the adverse environmental impacts likely to follow development for the original project design and any alternative approaches. Countermeasures are proposed. Further stakeholder participation may be desirable.

Step 6: Evaluation. Once the grounds for decision making have been set, an evaluation of the project can be completed from the information and analyses generated and a strategy for environmental management for the subject project can be prepared. This will include viable engineering alternatives and identification of designs, locations, precautions to prevent avoidable impacts, and recommended mitigation strategies.

Step 7: Reports. A main EIA report is prepared including recommendations to decision makers on whether and how to permit the project. Two additional reports are recommended: (i) environmental management plans, including countermeasures and mitigation (see Section 5.5); and (ii) baseline/monitoring requirements (real time and retrospective or post-audit).

Several techniques are available to help inexperienced assessors in the Screening phase. The most popular are: "checklists" which list all possible impacts and sometimes numerically rate their significance; and "matrices" which cross check various project activities (rows) against possible impact types (columns) and note the probable extent of impact.

The Assessor is a diagnostician and detective not just a data compiler. Environmental assessment should *not* be viewed as an exercise in data collection with a more-is-better approach. No data should be collected except that needed to examine particular hypotheses created to address specific impact potentials. The quality of EA/EIA is judged by the accuracy with which it identifies significant impacts and suggests practical countermeasures, not by how thick the report is.

6.1.2 SOCIAL IMPACT ASSESSMENT

Social Impact Assessment (SIA) is an addition to the EA process that is meant to explain the social consequences of environmental change. As addressed here, it is not meant to be a free standing social assessment process but rather an interpretation of the particular chain of impacts that starts with environmental change and leads thence to social change.

SIA is a way of trying to figure out what can and what does happen to people, their organizations, and their communities related to a particular environmental change. It involves the use of social science techniques to make predictions and to monitor results and evaluate outcomes. It is aimed at fair dealing with the various people affected by development. There is a modest literature on SIA techniques as they apply to CZM; as examples, see *Coastal Zone Management Handbook* [1] and "Social Science Techniques; Overview" [in Ref. 33].

6.1.3 ECONOMIC ASSESSMENT

Economic impacts can also be incorporated into the EA [32]. But the complexities of evaluating coastal renewable resources require that special methods be devised and used in CZM. This requirement applies to countries with political systems ranging from *laissez-faire* approaches to centrally planned economies.

Regardless of the type of economic system that is in place, a special mix of economic trade-off analyses is required for CZM because of the nature of coastal problems and its common property resources. First, many of the goods and services produced by these systems in their natural state are not easily expressed in cash terms, and second, many of these goods and services are harvested "offsite"; that is, they are external to the subject resource system and become economic externalities to adjoining systems. It is very important to separate market and non-market values as well as on-site and off-site locations of goods and services. Day *et al.* [34] comment that: "Traditional economic analyses generally recognize that, when externalities are large, market failure occurs . . ." and that with ". . . large externalities, the economic system, in effect, cannibalizes itself".

The absence of quantifiable market values for many environmental goods and services does not present an insurmountable problem. In addition to such standard forms of analysis as net present value, internal rate of return, benefit–cost ratio, and cost-effectiveness analysis, there are special analyses for situations where monetary values have to be measured indirectly, such as shadow pricing, willingness to pay, contingent valuation, and so forth.

Unfortunately, there are not simple "cookbook" techniques that can be employed in conservation economics. It will usually be necessary to engage a professional resource economist to assist with economic analyses.

6.2 *Rapid Rural Assessment*

Rapid Rural Assessment (RRA) is a process for gathering and analyzing information from and about rural communities in a brief time period (weeks). Sometimes called Participatory Rural Appraisal (PRA), the approach is somewhere between formal survey and unstructured interviewing. RRA sets out to create a dialogue with rural stakeholders [35].

RRA collects information by interview (group or single) on social values, opinions, and objectives and local knowledge as well as "hard" data on social, economic, agricultural, and ecological parameters. Mapping is the key to success. The value of the data produced depends largely on the collecting team's skill and judgement. A participatory approach, rather than a researcher–subject relationship, is the goal. Some of the advantages of PRA methods for learning the needs of communities are: they are cost-effective; sampling errors and bias are reduced; close discussion with rural people is enhanced; and there is flexibility to make adjustments during fieldwork [35].

Participatory mapping is particularly important, including: (i) social, such as village layout, infrastructure, population, households, chronic health cases, size of family; (ii) resources, such as fisheries, forests, land use, soil types, land and water management, watersheds, degraded resources, etc.; (iii) transects, such as walks and boat rides to see indigenous technology, resources, and fishing practices, to confirm resource mapping; and (iv) time links, such as local history, "time lines", and seasonality.

6.3 *Carrying capacity analysis*

Carrying capacity analysis was created (in the 1960s) as a method of prescribing the limits to development using numerical, computerized calculation with cold objectivity. It has not achieved much success in influencing government policy because of the complexity of the parameters and because politicians and administrators are reluctant to have their judgement pre-empted by a computer. Nevertheless, a non-prescriptive and more qualitative and normative concept of carrying capacity has been useful in influencing control of development, particularly tourism [36].

It is clear that resources are finite; they cannot withstand unlimited use. Already, in the late 1990s, many coral reefs are degraded, fisheries depleted, and beaches eroded away. The idea that there *is a limit* — a "carrying capacity" — has to be embraced to ensure that natural resources are not destroyed.

With carrying capacity, as with other biological analogies, human nature complicates the procedure for estimating limits. Some of the key components — such as tourist or user satisfaction — change when the users themselves or their preferences shift. Therefore, in spite of simulation models, the actual carrying capacity limit, in numbers of users or any other parameter, may be a judgement call based upon the level of change which can be accepted [36]. An analytic approach conditioned by semi-subjective factors can be useful (see Bonaire case, Section 8.5).

6.4 *Water quality control*

Water quality is a key parameter. CZM programs should focus on special coastal pollution sources that are not presently addressed by a national pollution agency, like those caused by watershed disturbance and runoff and, especially, those that affect Special Habitats. Of particular concern is sediment runoff from construction sites, farmlands, forest cutting, and land clearing operations. Also, pollutants flushed by storm runoff into coastal waters can create toxicity (biocides, oil wastes, etc.) and bring excessive nutrients (fertilizer, animal and human wastes, etc.) into coastal waters [37].

Control of land-based pollution is an important tool for CZM but it is very complex. It can require major changes in agricultural and industrial practices, as well as the development of advanced waste treatment technology. The main changes needed include: comprehensive control of sewage discharges; reduction of phosphate detergent use; reducing runoff of fertilizers and livestock wastes from farmlands through adoption of high standards of land husbandry; and limiting industrial effluents through more efficient use of resources [38].

6.5 *Land use control and zoning*

Because the unified CZM process operates at the interface between land and water, CZM may be put in the role of mediating between these conflicting water side and land side uses through land use allocation, or zoning. Space along the shoreline is in strong demand for human settlements, agriculture, trade, industry, amenity and marine transport activities such as shipping, fishboat harbors, and recreational marinas. This demand focuses on the water's edge, that is, the line of Mean High Water and the land and water that straddle that line. Because shorefront land has high value in both centrally planned and free market economies it is often necessary to allocate (zone) coastal land uses according to a broader social perspective than ordinary land.

Land use planning identifies parcels of land for particular classes of use, such as ports, warehouses, condominiums, houses, shops, nature reserves, open space, and other zones. This "zoning" process identifies areas most suitable for conservation as well as for development projects. Zones are delineated and regulated to accommodate prescribed uses. CZM land use planning requires extensive socio-economic and spatial/land information as does traditional land use planning.

Population, commerce, and spin-off industries are often attracted to the Coastal Zone by the development of large factories and either may be environmentally damaging, imposing higher costs on the community for streets, police, fire protection, schools, and other essential services. Therefore, land use planning and zoning decisions include the secondary development that industry will induce.

CZM land use adjustment helps to preserve especially productive or scenic natural resources, such as national scenic areas or national parks. The coastal landscape is special and may need special CZM attention to protect its scenic quality and to guarantee people's access to beaches and waters.

6.6 Setbacks

The ocean beachfront is a most hazardous place to build a factory, warehouse, hotel, house, or store. Continued severe beach recession is certain and predictable along much of the coastline of many countries. It is unwise to allow development of property that will certainly be lost to the sea due to future erosion, especially when the security of buildings so often creates demands for groins, bulkheads, and other protective works, which may further imperil the whole beach system. Erosion is quite predictable and the risk should be obvious to those who build there (see Anguilla case, Section 8.1).

A key component of unified CZM is a "setback" provision whereby coastal development is kept back a safe distance from the shoreline (Fig. 9). This not only protects structures from storm waves and erosion, but it conserves the natural defenses of the shoreline, e.g. sand-dunes and mangrove forests, which otherwise would be bulldozed, cleared, and filled (see Section 4.7.9). The same setback that protects beachfront structures from erosion and storm waves can also preserve turtle nesting sites on the back beach [24].

A setback line is delineated at a calculated distance inland from the beach and all construction is required to be located landward of this line, following these steps.

1 Predict how far back the beach will erode in the future (say 50 years from now)—historic aerial photo series will help in predicting long-term rates of erosion.

2 Identify this setback line on appropriate land use planning maps.

3 Prohibit any building, or rebuilding, seaward of this line.

In this way setback lines provide vacant, open space strips between the ocean and the closest permanent buildings. In a CZM program, setbacks may form a specified "line of retreat" which not only keeps future structures from locating close to an eroding beach, but attempts to restrict any expansion or rebuilding of any of the existing structures that are seaward of the setback.

There are other uses for setbacks; for example a "buffer strip" around the edge of a mangrove area to prevent encroachment. Or a setback for storm protection—a rule of thumb is that a mangrove strip of at least 50 meters wide is needed for attenuation of secondary storm waves, e.g. a wave 1 meter high [16]. Other examples are setbacks to prevent buildings from blocking the coastal viewscape or for preventing pollution (as from shoreline garbage dumps). In some cases the coastal setback protects a "green belt"—a band of mangrove and other trees planted along the shoreline for storm protection (as has been done in Bangladesh).

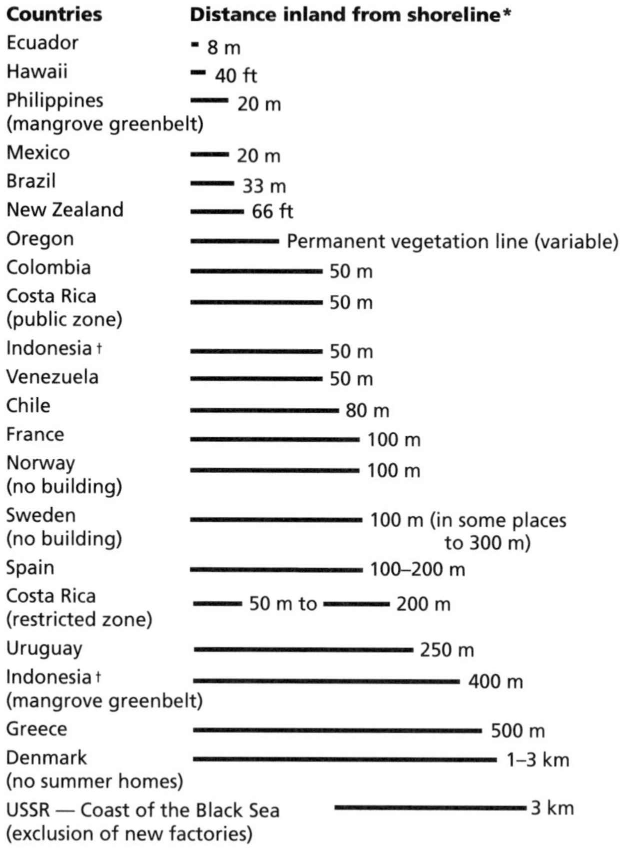

Countries	Distance inland from shoreline*
Ecuador	▪ 8 m
Hawaii	▬ 40 ft
Philippines (mangrove greenbelt)	▬ 20 m
Mexico	▬ 20 m
Brazil	▬ 33 m
New Zealand	▬ 66 ft
Oregon	▬▬▬▬ Permanent vegetation line (variable)
Colombia	▬▬▬▬ 50 m
Costa Rica (public zone)	▬▬▬▬ 50 m
Indonesia †	▬▬▬▬ 50 m
Venezuela	▬▬▬▬ 50 m
Chile	▬▬▬▬ 80 m
France	▬▬▬▬ 100 m
Norway (no building)	▬▬▬▬ 100 m
Sweden (no building)	▬▬▬▬ 100 m (in some places to 300 m)
Spain	▬▬▬▬ 100–200 m
Costa Rica (restricted zone)	▬ 50 m to ▬ 200 m
Uruguay	▬▬▬▬ 250 m
Indonesia † (mangrove greenbelt)	▬▬▬▬ 400 m
Greece	▬▬▬▬ 500 m
Denmark (no summer homes)	▬▬▬▬ 1–3 km
USSR — Coast of the Black Sea (exclusion of new factories)	▬▬▬▬ 3 km

Fig. 9 Setbacks used by various countries to hold structures back from the shoreline. * Definition of shoreline varies, but it is usually the mean high tide. Most nations and states exempt coastal dependent installations such as harbor developments and marinas. † Indonesia has both a 50 meter setback for forest cutting and a 400 meter "greenbelt" for fishery support purposes. (Source: Sorensen & McCreary [20].)

6.7 *Special Habitats*

A combination of regulatory and custodial (ownership) approaches provides optimum protection of Special Habitats because it includes both a regulatory scheme for conservation and a program for establishing Resource Reserves (conservation) and marine national parks (recreation, education).

6.7.1 PURPOSES

There are two main purposes for identifying special habitats and providing for their protection: conservation of the economic resource base (fisheries, tourism, etc.) and preservation of biodiversity (the whole range of species and natural habitats).

Of particular importance in biodiversity conservation are the habitats of species that have been designated as especially valuable or in danger of extinction [38].

6.7.2 IDENTIFICATION

CZM should recognize three types of particularly valuable habitats by identifying the following three categories.

1 *Generic types of habitats.* Those that are widely recognized as highly valuable and that should be given a high degree of protection through *regulatory* mechanisms — wetlands, seagrass meadows, coral reefs, species nesting sites. All should be mapped and publicized. In the CZM process of project review, developers would be required to avoid these types of habitats; therefore, developers must be informed ahead of time (before they design projects) that restrictions exist. In addition to ecologically valuable areas, other types of areas should be identified, such as sand-dunes (which stabilize beaches) and flood-prone lowlands (those that are regularly flooded) both of which would be included in a "natural hazards prevention" category.

2 *Specific sites.* Those that are identified as Special Habitats and should be identified for *special regulatory protection.* These would include certain specific (named) lagoons, estuaries, islands, mangrove forests, river deltas, coral reefs, and so forth. Each would be described, mapped, and announced for the knowledge of all interested parties. The CZM authority would strongly constrain development in these site specific habitats by regulation. As "red flag" areas, they would get special analysis in the development review process.

3 *Resource Reserves* and other protected areas like marine national parks would include Special Habitats and critical resource areas that need the additional safeguard of the type of *custodial protection* through ownership that is awarded to terrestrial parks and reserves. These might best be assigned to the country's existing conservation agency. Proprietorship generally confers a higher level of autonomy than does regulation through the "police power".

As a real world example, the unified CZM system for Puerto Rico (US Territory) is close to the above, as shown below.

1 *Generic habitat types.* All mangrove forests are included in a "special planning areas" category.

2 *Specific sites (critical areas).* Numerous bays, lagoons, and other coastal features are included in a list of site specific special planning areas.

3 *Protected Areas.* Numerous coastal areas of exceptional natural value are identified as potential Natural Reserves.

6.7.3 RESERVES AND MARINE PARKS

Resource Reserves, as mentioned above, support the broad objectives of CZM by: conserving special nurturing areas for fish species, enhancing tourism revenues and recreational benefits, preserving biodiversity, promoting baseline scientific studies, etc. A coral reef reserve could be established to both conserve the reef habitat and protect the beach from wave attack during storms.

Heyman [39] made an intensive study of the subject of protected areas and identified two main categories: "reserve-like" areas and "park-like" areas. The first category is aimed mainly at conservation and the second mainly at recreation and education. But both are based on ownership of the property to be included in the protected area, which is simpler for the commons of the sea (already under government control) than for privately owned areas.

Zoning is a major tool for reserve and marine park management. With zoning the multiple-use proposition for the whole reserve is enhanced through subdividing it into single use areas such as: fishing, water sports, ecological reference areas (controls), stock replenishment (sanctuaries), diving/snorkeling, aquaculture, etc.

Economic considerations are essential. Resource Reserves need prohibit only those uses which are incompatible with the primary purpose of the area [27]. Multiple use is the key — the greatest variety of compatible economic uses should be permitted, those that neither deplete nor significantly disturb the resource.

In the evolution of protected area approaches, some holdings now resemble "resource management areas" more than "protected areas". These management areas are run by a "management authority" of some type rather than a wildlife or national park type agency and are organized to yield a wide variety of uses and thus are really "multiple use" areas; examples are the Great Barrier Reef of Australia (see Australia case, Section 8.2) and the Florida Keys Coral Tract of the United States. It is important that such areas be closely coordinated with the CZM program.

Note that design and management of protected areas is a complex subject, one that whole books have been written about, e.g. the IUCN book *Marine and Coastal Protected Areas* [37] and the Taylor and Francis book *Managing Marine Environments* [40].

6.8 Restoration

Ecosystem restoration is an important objective of unified CZM. The CZM planning unit should survey and identify the Special Habitats that have been degraded and that can be repaired at reasonable cost and effort. These should be mapped, priorities assigned, and strategies for rehabilitation created.

While all coastal resources that have been degraded cannot, in a practical sense, be returned to full productivity, some of them can. Mangrove forests can be replanted, coral reefs can be started toward gradual renewal, and normal circulation to wetlands can be restored.

If a wetland is covered with fill behind a concrete bulkhead, it would be unrealistic to plan to restore it to its original condition; but if a wetland has been diked for rice culture or aquaculture, it would be relatively easy to remove dikes, restore circulation, and reconvert it to a nearly natural wetlands condition.

If a coral reef has been damaged by pollution, hurricanes, mining, or boat anchoring, it would be difficult, but not impossible to rehabilitate it. Such rehabilitation can be costly and the time of recovery very long, but for certain reefs of high value for tourism, fish breeding, or shore protection, the investment could yield a high payoff [41].

Hundreds of thousands of hectares of mangrove have been planted in restoration and shore protection initiatives, globally. There have been great successes and disappointing failures but on the whole the projects have been successful. A clear message has arisen from these initiatives: plantings should not be wasted in environments that would not *naturally* be colonized by mangrove.

Community based restoration projects may be far less expensive than those done by hired labor; for example, in a Philippines analysis, community-based mangrove plantings cost about US$80/hectare while contractor plantings cost more than US$400/hectare. The new mangrove area is also better cared for when the community plants it and has special rights to it.

Degraded dunes and beaches can be rebuilt using the approach of the "sand budget" [16] whereby one treats the beach as one might treat a bank account in that there are inputs and outputs, credits and debits, along with cash reserves. The parameters are as follows:

1   *Debits.* Longshore downdrift transport, wind transport out of area, offshore transport, deposit in submarine gullies, mining, aeolian (wind) transport, and solution and abrasion (withdrawals).

2   *Credits.* Longshore transport onto beach, river transport, sea cliff erosion, onshore transport, biogeneous deposition, hydrogenous deposition, and beachfill (deposits).

3   *Sand storage.* Dunes, berms, sandbars, offshore sinks (reserves).

6.9 *Survey*

An important need for CZM is identifying, evaluating, delineating and mapping habitats to be given conservation status, such as mangroves, coral reefs, dune fields, seagrass meadows, and species habitats. Hazardous areas such as coastal flood plains and high-hazard zones should also be described and delineated. Polluted areas should be located and mapped. Some components of survey work should not be postponed because the results are needed for strategy planning. Other components could be done for Master Plan preparation. Some secondary data may already be available on computers or in libraries.

Another important survey job is identifying, evaluating, and delineating the following: (i) specific environmentally critical areas to be red-flagged for regulatory protection through the development project review process; and (ii) special areas to be recommended for reserves, sanctuaries, national parks, refuges, or other protected areas status.

6.10 *Information techniques*

Improving the coastal data base is required by CZM in most instances, including collection, storage, analysis, and reporting aspects (see Section 4.7.14).

6.10.1 MAPPING

It is expeditious to organize the data base so that essential information can be mapped and also to display as many categories of data as possible on maps, using Geographic Information Systems (GIS) or other approaches. The first step in information gathering for CZM should be preparation of good base maps at appropriate scales. For example, Goeghegan *et al.* [21] state: "It has been found time and again that perhaps the most useful way for the environmental planner to discover trends, conflicts, and problem areas that can otherwise be easily overlooked, is by mapping information".

An interesting example is a Situation Management initiative in St. Lucia, about which Renard [42] commented that the preparation of a map of marine resources, issues, and conflicts made by a mix of participants (during a boat inspection of the area) was "particularly important" to the success of the program, not only because it produced valuable data but also because it drew the data providers (fishers, divers, scientists, managers) together and "credentialled" them to the other participants including government officials.

6.10.2 REMOTE SENSING

Remote sensing—by satellites such as LANDSAT (USA) and SPOT (France) —can be helpful, but the images are often very expensive. The high spatial resolution of multiband radiometers on LANDSAT and SPOT, well proved for landside survey, also works well for shallow water survey. Remote data have their best use in coastal zone planning and management when coupled to digital mapping and GIS technology. A new US satellite, LANDSAT 7, features an advanced Thematic Mapper and the output will be inexpensive (contact Frank Muller: carib@carbon.marine.usf.edu).

6.10.3 GEOGRAPHIC INFORMATION SYSTEMS

Computer-assisted mapping tools, used in storing, retrieving, processing, and displaying spatial data may be particularly useful. The most popular of these are GIS. Previously requiring large computers, GIS are now available on microcomputers and work stations [43]. This increasingly puts them within the budget of most institutions which deal with resource data (Fig. 10).

GIS computers can be programmed for direct production of resource maps. Most coastal features that have spatial attributes can be stored, analyzed, and printed out as maps using simple electronic methods; for example, the distribution of fisheries resources in coastal waters can be mapped. Decisions in resource use often depend on the spatial distribution of the resource in relation to other factors such as transportation. FAO has prepared a useful guide, *Marine Resource Mapping: An Introductory Manual* [44].

Most resource data with spatial information are readily inputted as points, lines, and areas, with attributes tagged onto these entities. The analy-

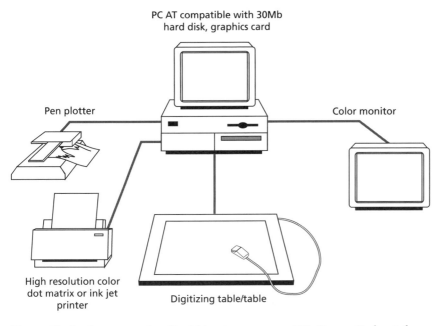

Fig. 10 The hardware setup for affordable micro-computer GIS. (Source: Butler *et al.* [44].)

ses for GIS are transformations of geographical data and attributes in the form of a map or referenced to a map. For example, GIS can overlay two map layers showing the extent of mangrove forests of two different years within the same study area and show exactly where mangroves forests have increased or have shrunk. Automatic calculation of area is another feature of most GIS [44].

The role of the GIS as a tool in processing and displaying resource data is unending [44]. Also, GIS are open ended and can easily receive new data or change old data. Therefore, the GIS data base can easily be updated. Then too, GIS systems work well in conjunction with remote sensing, including satellite images. The major difficulty is acquiring the data by which to do this.

6.10.4 BASELINE AND MONITORING

When approving a coastal development project with potential environmental impacts, the CZM approach involves baseline data and monitoring to evaluate actual impacts. This is often designed as part of the EA process. The baseline is the "benchmark" which represents the current status of habitats, water quality, and other relevant indicators. The baseline data set should be designed to measure current ecological conditions at those specific points where the project is expected to have impacts. The baseline should be kept as simple as possible. Monitoring activity measures any future change against the baseline condition. Data collection will most often be done by consultants. Two important kinds of monitoring are:

1 The *tactical level* is the "oversight", or "real time", monitoring which is done in conjunction with the project construction phase. It should be specified in the environmental management plan (EPM) for the project; the object is to monitor the construction operation day by day so as to detect any major negative impacts that may be occurring.

2 The *strategic level* is the "retrospective" (hindsight) monitoring which is done to compare measurements of certain key characteristics of the environment before and after a project so that a determination can be made of the project's effects. This of course requires that "benchmark" data be collected before the project starts to provide statistically sound before-the-project baseline information to compare to after-the-project information.

6.11 *Conflict resolution*

In most coastal countries there is an array of conflicts due to negative interaction among user groups in congested marine areas [29]. The coastal sea generates conflicts related to transport, fishing, dumping, mining, oil extraction, and so forth. Coastal land is used for human settlement, agriculture, trade, industry, and amenity, all of which are potential areas of conflict.

Multiple use, while a good objective, can lead to competition and conflict. A major benefit of the unified CZM approach (integrated, multiple-use oriented) over the traditional sectoral (single use) approach is that it provides a framework for resolution of arguments over who gets to exploit which coastal resources and how and when.

Conflicts that arise over coastal environmental issues can be severe, including loss of life (e.g. the Indonesia fishing "wars" of the 1970s), and destruction of property (e.g. a riot of 50 000 people who burned down a tantalum factory being built in Phuket, Thailand, in 1986 because the community was not consulted [1]). In fact, it is intense conflict over use of the coast that so often makes the CZM process appropriate.

Objective methods for resolving conflicts include: fact finding and executive decision, study commissions, bargaining sessions, informal negotiation, formal mediation through facilitated dialog, administrative or public hearings, and adjudication. Any method can be included in a CZM framework.

The most effective method is often mediation, using *facilitated dialog* in which developers, industry, commerce, environmental groups, communities, and local and central government agencies are typically represented in the negotiations. The mediating/coordinating entity must look at all sectors with legitimate interests to find the most broadly compatible solutions [1,20].

7: Coastal Connections

*Behold how good and how pleasant it is for brethren to
dwell together in unity.* [Old Testament, Psalm 133]

The key to Coastal Zone Management (CZM) is unity of different stakeholder interests and integration of government agencies and economic
sectors. The process of establishing goals and objectives, as well as the
process for reaching consensus on the set of issues the program should
address, can be facilitated by the stakeholders agreeing on a *shared vision* for
the future of their coastal community.

7.1 *Unification*

Governmentally, coastal zones and their resource systems are complex
because of the need to counter mixed jurisdiction, dispersal of authority,
multiple user conflict, and extent of common property resources involved.
But unified CZM programs are designed to handle such complexity by creating a framework for collaboration, integration, and communication between
economic sectors in a country and its communities and between agencies
that guide those sectors and assist the communities.

What unification does is to ensure lesser emphasis on single sectoral concerns and individual demands and more emphasis on a collective agenda
under which all parties participate. Participation serves to unite people in the
sharing of needs and ideas and in the working of solutions. Therefore it is
necessary in the policy formulation and planning stages to consult with all
stakeholders.

Unification requires connections among government agencies, various
economic sectors, politicians, coastal communities, non-government organizations (NGOs), military, academics, and other stakeholders. Also important are community based (self-reliance) programs, formation of coalitions,
effective communications and education programs, and clarifying the moral
basis for sharing resources with neighbors and future generations.

7.2 *Coordination*

The job of coordination is complex and important because all levels of government must be involved in CZM along with all development interests and
resource users. It requires efficient communication and effective dialog.
Information sharing is especially important. Public hearings or consultations
may be appropriate in advance of particularly important decisions. The com-

plexity of this endeavor is illustrated by the multitude of agencies and issues shown in Table 1.

The CZM dialog should include private sector and military spokesmen as well as agencies and NGO groups. The CZM process thus becomes a cooperative venture incorporating the efforts of the economic sectors, government agencies and all private stakeholders including researchers, extension workers, rural bankers, community leaders, and NGOs.

7.2.1 INTERGOVERNMENTAL COORDINATION

Unified CZM is based upon coordination among all levels from national to village governments — this is "vertical integration". At one extreme, local governments are involved because they govern where development takes place, where resources are found, and where the benefits or disbenefits are mainly to be felt. Local level activities must be endorsed and supported by the higher levels of government [46]. At the other extreme, central government has to be involved because responsibility and authority for marine affairs inevitably rests there—navigation, national security, migratory fish, international relations, etc. (see Section 4.6.1). At the intermediate level, provincial or regional governments are involved because they also often have financial and administrative responsibility in the coastal area (see Section 4.6.2). The need is for a power sharing mechanism.

"Horizontal integration" refers to cooperation among *competing* agencies to assure integration and unify and coordinate the separate economic and governmental sectors and thereby to reduce fragmentation and duplication [20]. The governance arrangement must have both horizontal and vertical integration.

For most programs it will be desirable, if not essential, to establish an "interagency coordinating committee" or "working group" (perhaps 10 or 15 persons) to review progress, consider program changes, discuss proposed new rules, receive advice, and consider actions on specific development applications and resource management proposals. This group includes spokesmen from local and regional entities. Staff from existing agencies can be assigned to this team on a part-time basis without undermining their normal work.

7.2.2 INTERSECTORAL COORDINATION

A distinctive feature of a unified CZM program is the fact that it is cross-sectoral and that it seeks to integrate or coordinate activities of a broad spectrum of existing private enterprise along with government agencies. It thereby ensures an appropriate shift from single sectoral concerns and self-centered concerns to a collective agenda.

Table 1 Generic responsibilities for government agencies on particular issues. (Source: Crawford *et al.* [45].)

Gov. departments: Issues:	Ministry Defense	Hydr. Service	Ministry Me. mar.	Harbor Office	Ministry Environ.	Ministry P. works	Ministry Health	USL	Ministry Culture	Ministry Industry	Region	Province	Com.
Navigation	R	M	R, P	Mg									
Fishing			R, P	M									
Ind. & com. harbors			R, P	Mg					R		P		
Marinas			P	R		C			R				
Water quality			M		R, M		R, M				P		P, Mg
Rural urban waste					R		R	M			P, R	M	Mg, C
Industrial waste					Mg		R	M			P		C, Mg
Eutrophication			M	M	P, M, Mg			M					
Coastal erosion					M	P, C					P		C, Mg
Tourism	R		R	Mg					M		P		R, Mg
Urbanization									M		R, P	M	C, Mg
Parks and reserves			P, R		P, R, Mg						Mg		Mg, M
Archaeological sites									R		P	R	Mg, M
Military uses	P, R, C, Mg	P, Mg, M											
Mapping				M									
Offshore activities			R, Mg							P, R			

Abbreviations of functional responsibilities: P = programming, R = regulation, Mg = management, M = monitoring, C = construction.

Coastal communities and their spokesmen must be directly consulted about the formation of new coastal policies and rules on resource use if they are to support them. According to Renard [47], public consultation is an opportunity available to the entire management community to ensure the quality and the effectiveness of the management solutions that will be implemented. He emphasizes that community involvement is also a *duty* because "... the issue remains, above all, one of human development" and because "... people are not the object of that development but the subject of development and the makers of their own history".

Encouragement of public participation is not supposed to lead to predetermined outcomes nor to change the ideologies or views of the fishermen, the government officials, the planners or the consumers. Nor is it supposed to be a means to get a particular group or sector "aligned" to the needs of another group. The purposes of such participation are to unite people in open discussion and sharing of needs and ideas and in the working of solutions (see Philippines case, Section 8.13). Public participation should lead to true consultation with ideas growing in both directions. Planners and managers too often resort to public consultation and involvement only when they encounter some form of opposition.

Kelleher [48] states that "... participation that is not actively encouraged is not real participation. You've got to go out there. You've got to go through the process of distrust before you get to the process of trust. You've got to be prepared to be insulted, contradicted, even threatened. You have to prepare yourself psychologically."

According to White [49], personal and community involvement come from wanting to support common values to gain some real or perceived benefit for the individual and the community. Without it marine resources can never be conserved nor sustained, because external enforcement of laws in the marine commons is not usually practicable.

7.3 *Education and outreach*

Civic awareness plays a major role in the success of CZM. In countries where CZM is effective, conservation awareness is usually high among communities, managers, *and* the private sector. The most important goal is to explain to the people the long-term, sustainable benefits that conservation can provide, through public information and education. Honest efforts to inform the public are essential—education should not be used as propaganda for "selling" CZM programs.

Environmental education aims to provide the community with information and a conservation ethic so that its members can make informed decisions about the use of their resources [29]. Then too, the place where the land ends is also the place where the knowledge and experience of most administrators ends. Planners, managers, engineers, and politicians alike need to be informed about the sea and the sea coast.

The first step in designing an education program for capacity building is to identify the main audiences; for example, artisanal fishermen, dive operators, tourists, hotel owners, port directors, and politicians. Because most politicians monitor their constituencies, public awareness and sensitization is important to success of CZM. In educating politicians and economic planners, it is important to use familiar language and concepts.

Initially, a multifaceted approach, combining printed materials, audiovisual presentations, and face-to-face interaction, is probably the best way to start an education program. Depending on the target audience and budget, a variety of additional options can be employed: mass media (press, television, radio), fixed exhibits, tours, training workshops, the sale of promotional items such as T-shirts, and informal recreational activities with an educational focus [29].

7.4 *Collaborative management*

Unified CZM requires power sharing and decentralization of authority (see Section 4.6). The product of decentralization of resource management has been variously termed "community based management", "joint management", "the partnership", "collaborative management", or "co-management". The movement, whatever its title, has been very popular with international workers during the 1990s as ". . . an essential feature of the emerging face of conservation" [50]. In short, collaborative management works because empowering communities always works better than commanding them.

Collaborative management requires networking, forging links with community leaders, local law enforcement officers, private business, and national agencies like tourist authorities and environmental and fishery agencies. The Situation Management approach to CZM is ready made for the collaborative approach in that it targets a specific coastal situation that needs management (see Ecuador case, Section 8.7). It combines normative, customary, scientific and legal approaches tailored to conservation at particular sites and it works well at coordinating local government and central activities.

Any move toward a democratic approach to CZM must itself be commended, but there are more than socio-political advantages to be gained. Most importantly, where a community has management responsibility, more care will be exercised in the use of resources; for example, the quantity of fish or shellfish removed will be controlled, abstinence may be practiced during spawning periods, and less destructive fishing methods may be used. Also, there may be a greater willingness to curb pollution and conserve Special Habitats.

The effectiveness of traditional, or customary, conservation such as fisheries management at the community level, has been recognized in studies in Nova Scotia (Canada), Brazil, Palau, the Solomon Islands, and elsewhere.

7.5 *International*

Because of the nature of the Coastal Zone, many of the coastal issues facing countries are of international scope. The list of transnational issues that require international cooperation includes: shared stocks, endangered species, highly migratory species and transboundary "straddling" fisheries, maritime transport, maritime boundary resolution, straddling oil and gas deposits, and transboundary pollution. There is a connection between national CZM programs and current international initiatives on biodiversity conservation, marine debris and entanglements, endangered species, hazardous wastes, wetland protection, and global warming. Thus, in planning coastal seas activities, it will be necessary to consider effects on adjacent countries, a function that can be served by CZM.

FAO has the responsibility of organizing fisheries conservation at the regional level. For pollution, the Montreal Guidelines for the Protection of the Marine Environment against Pollution from Land-based Sources, drafted by the United Nations Environment Program, is involved along

Fig. 11 French soldiers cleaning the beach at the Brittany coast after the *Amoco Cadiz* grounding. (Photo by J.R. Clark.)

with the World Health Organization, and United Nations Development Program (UNDP's) East Asian pollution programme, MARPOL.

Several international agreements bear on coastal conservation matters. For protection of species and their habitats, relevant instruments include: the Ramsar Convention which is an international treaty for protection of wetlands; Control of International Trade in Endangered Species (CITES) which restricts trade in either live animals or their products; and the Regional Seas program of UNEP which develops protocols meant to conserve sealife and ecosystems. The International Maritime Organization (IMO) is the parent of a convention on emergency preparedness and response (Fig. 11) and is creating routing measures for tankers along with identifying "Particularly Sensitive Sea Areas".

The economic development financing agencies (e.g. World Bank, USAID, EEC, Inter-American Development Bank, and Asian Development Bank) realizing the environmental limitations to development have required environmental controls on funded projects and are supporting primary environmental projects. Fortunately, the client countries themselves are showing more interest in sustainable development and related issues but are concerned with food security and looking for an equitable sharing of the costs of world environmental protection.

8: Acts of Stewardship

It is He who has made the sea subject, that ye may eat thereof flesh that is fresh and tender, and that ye may extract therefrom ornaments to wear. [The Koran, Sura 16, aya 14]

The mini case studies of coastal management initiatives presented in this chapter cover a range of experience from single objective to unified multi-objective coastal management actions. They are only a sample of the numerous experiences with Coastal Zone Management (CZM) that could be cited. Some are miniaturized updates of cases that appeared in the *Coastal Zone Management Handbook* [1].

The cases are mainly from developing tropical countries. Because case authors are responsible for provision of information, literature citations are not given. Each mini case has been edited to conform to requirements of this book; the author accepts full responsibility for any errors.

8.1 *Anguilla: management of the beachfront*

The sand beach is a most valuable natural and economic resource to the small islands of the Eastern Caribbean, where beach-oriented tourism is often the main industry. All too often, tourism structures and facilities are positioned too close to the beach, as are houses, roads, and airports.

In Anguilla, as in most Eastern Caribbean Islands, planners have tried to cope with erosion by designating fixed setbacks for all beaches. However, Hurricane Luis in September 1995 has proved that these setbacks are often too narrow and are inadequate to either conserve beaches and dunes or protect coastal infrastructure. In Anguilla, it was observed that on average, the beaches and dunes had retreated 9 meters inland during the hurricane and the damage to coastal infrastructure was extensive.

Following Hurricane Luis, the Government of Anguilla obtained assistance from the British Dependent Territories Regional Secretariat to undertake an assessment of the problem. One of the studies created a methodology for setbacks which would reduce future damage from hurricanes [14]. Coastal development setbacks are especially important in tourism-oriented islands because they do the following.

1 Provide buffer zones so that beaches may move naturally without the necessity for sea walls and other structures.

2 Reduce damage to property during high wave events like hurricanes.

3 Provide improved vistas and access along the beach.

4 Provide privacy for resort visitors.

The use of a fixed setback for all beaches has proved difficult to implement politically because of developer resistance. Too often, too little has been done to educate the public or government administrators about the purposes of setbacks. As a result they are seen as constricting to development and unnecessary.

Beaches behave in different ways; some are eroding while others are accreting, and not predictably. As a result, the guidelines developed for Anguilla sought to develop specific setbacks for individual beaches based on a combination of different parameters.

1   Historical changes over the last 30 years.

2   Recent changes over the last 5 years as determined from beach monitoring data.

3   Changes in the position of the dune line expected from a category IV hurricane such as Hurricane Luis.

4   Coastal retreat likely to occur as a result of projected sea level rise over the next 30 years.

5   The existence or absence of offshore features such as coral reefs and beachrock ledges which provide protection during high wave events.

6   Anthropogenic factors such as beach and dune sand mining.

7   Planning considerations, e.g. lot sizes, marine park designations, and special development features such as "beach bars".

The actual setback for a specific beach was determined as follows:

$$\text{setback} = \left(p\right) + \left(h\right) + \left(s\right)$$

where p is the change in coastline position (based on historical and recent changes); h is the change in position of the dune line/coastline resulting from a major hurricane; and s is the change in position of the coastline resulting from predicted sea level rise over the next 30 years.

Once the setback was calculated, offshore features, anthropogenic factors and planning considerations were subjectively included in the calculation where necessary by multiplying the setback value by an appropriate factor. Using this methodology, specific setbacks were determined for each beach. In all cases, setbacks are measured from the line of permanent vegetation.

In Anguilla the setback values range from 18 meters to 92 meters. Aside from the beaches, "blanket" setbacks were determined for other coastal types existing in Anguilla as follows:

1   15 meters from the cliff edge (cliffs in Anguilla are made of limestone);

2   30 meters from the natural vegetation line on low rocky shores; and

3   on small sandy offshore cays, if development is permitted, it is to be of a temporary nature with wooden piles and all wooden construction.

These setback guidelines have been incorporated into the Draft National Land Use Plan, which has been submitted to the Executive Council for its approval. In the meantime, the Land Development Control Committee is implementing the guidelines.

The setbacks determined by this methodology allow for geographical variations between beaches and can be fully justified and explained to developers, but nevertheless, continuing education is essential for the public, and for special interest groups such as architects, contractors, and politicians. Therefore, the Physical Planning Department engaged in creating environmental awareness including radio talks and discussions, slide presentations and distribution of leaflets along with a regular newsletter to the general public and the Executive Council. Unfortunately, some developers have the influence to appeal to the Executive Council and are permitted to build as little as 5 meters from the line of permanent vegetation.

At Anguilla we learned that flexibility is needed by planners to deal with developers and cope with socio-economic factors. However, interpretation of the coastal setback guidelines should not be so flexible that they lose force. The Anguilla case has also shown that new coastal management practices can often be tested by putting them into practice on an "informal basis", which allows for review and fine tuning to find out what really works before the law or regulation is passed. Finally, it may take a major disaster, such as a hurricane, to provide the impetus for changing management practices.

Contributed by: G. Cambers, University of Puerto Rico, Sea Grant College Program and Orris Proctor, Physical Planning Department, Anguilla, W.I.

8.2 *Australia: management authority for the Great Barrier Reef*

The Great Barrier Reef is one of the most diverse ecosystems on earth. Spreading over a distance of more than 2000 kilometers along Australia's north-east coast, the Great Barrier Reef comprises 2900 coral reefs, a vast range of inter-reefal habitats, and about 1000 islands.

The Great Barrier Reef Marine Park is one of Australia's richest assets, supporting a wealth of recreational activities, extensive tourist industries, long established fisheries, a major shipping route, jobs for coastal people, and of inestimable value to Aboriginals and Torres Straight islanders.

During the late 1960s and 1970s concern was raised about the changing density of human use of the Great Barrier Reef. Of major concern were proposals for oil drilling and limestone mining. Other concerns were increased land clearing and development along the adjacent coast as well as accelerated fishing, recreation, and tourism.

To address these concerns, Federal Parliament acted to establish the Great Barrier Reef Marine Park Authority in 1975. The Park provides for multiple use consistent with requirements for nature conservation. The Act itself banned oil drilling and mining as unacceptable threats to the coral ecosystem.

Zoning plans, and more detailed management plans for extensively used areas provide the basic framework for management of the Park. The focus of the Authority has changed over time. For the first 10 or so years the focus was on establishing the Park and management systems. Now the focus has shifted to resolving the following major issues: (i) maintenance of water quality; (ii) effects of fishing; and (iii) effects of shipping, and maintaining biodiversity. Awaiting intensive effort is full integration into the program of the needs of Aboriginal people whose lifestyle and culture have evolved over thousands of years of co-existence.

A summary of important lessons learned from the Authority's experience include the importance of:

1   adopting an holistic approach to ecosystem management;
2   establishing an independent authority with strong legislative mandate to focus exclusively on management of the protected area;
3   establishing formal complementary management arrangements amongst all relevant levels of government and stakeholders and creating processes for reaching agreement on proposed restrictions;
4   not postponing decisions awaiting perfect information but using the best available scientific information and the precautionary principle;
5   gaining the support of affected communities and involving them in the decision making process;
6   providing adequate funding both to the management authority and to supporting agencies in accordance with formal agreements.

Contributed by: P. McGinnity, Great Barrier Reef Marine Park Authority, Townsville, Queensland, Australia.

8.3 *Australia, Port Phillip Bay: a failed attempt*

Port Phillip Bay in south-eastern Australia, with Melbourne on its northern shore and Geelong on the south-west, has a surface area of about 1920 km². The coastline is varied, with some cliffs, extensive beaches, and small areas of mangrove and salt marsh. About half the coastal fringe is preempted by government agencies for port operations, navigational facilities, military purposes, and wastewater treatment. Remaining areas are used as reserves for public uses, including facilities for boating, car parks, picnic areas, and nature reserves. It was clear that these uses required CZM-type management to avoid conflicts and maintain the environmental quality and scenic aspects of the Bay. In 1966 a CZM initiative was commenced but after decades of political activity, public debate, and media involvement, it failed.

The Port Phillip Authority was allotted the zone extending 200 meters landward and 600 meters seaward from the low tide line. It was multiple use oriented with a mandate to assist the Victoria government. The Authority consisted of four government departments with a Consultative Committee including other agencies, local governments and the general public. A permit was required for all structures in the Zone. Recognizing a Zone of Influence, the Authority also consulted hinterland and offshore management agencies over such problems as pollution from rivers and from ships at sea.

A typical issue was the proliferation of artificial shoreline structures, such as sea walls, groins, and breakwaters, designed to combat coastal erosion and provide mooring facilities for boats of various kinds. Depletion of adjacent beaches, a major problem, was partly offset by artificial beach restoration.

The scope of the Authority was impeded by limited funding and by existing agencies' unwillingness to share authority. Soon the Authority became essentially a permit-issuing agency, was branded an unnecessary bureaucracy, and then disbanded in 1980.

By default, coastal management was put in the hands of three agencies: the Ministry of Planning and Environment, the Department of Conservation, and the Melbourne Port Authority, interacting with other government agencies. Eighteen local governments formed 39 committees for management.

Among the lessons learned is that interagency rivalry is the burden that all CZM programs bear. The Port Phillip Authority failed largely because of this rivalry and other politics. Management reverted to the old-style political resolution of competing demands by agencies with inevitable disputes.

In general, conservation groups and the public often fail to support CZM of the kind the Authority had created. Without CZM, decisions result more from political expediency, electoral manoeuvring, and media pressure than from any scientific or economic logic.

Source: Adapted from E.C.F. Bird in *Coastal Zone Management Handbook* [1].

8.4 *Barbados: incremental CZM*

The economy of Barbados, a small Caribbean Island of 450 km² and a population of 250 000, is based primarily on tourism that is so dependent on coastal resources ("sun, sea, and sand"), that more than 90% of resorts are located along the coast.

The challenge in Barbados is not only to protect the coastal infrastructure, but also to conserve the beaches which are so important to the tourist industry. Up until the 1950s the Coastal Zone was sparsely inhabited because of the prevalence of storms and flooding by the sea. However, with the advent of tourism in the 1960s, the coastal swamps were cleared and developed with hotels and residences.

During the 1970s, coastal erosion, involving human-induced factors, became a serious problem. Most of the coastline on the west and south coasts had become permanently anchored by hotels and other infrastructure; thus as the erosion continued, the beaches narrowed and steepened, and in several instances the hotels themselves were endangered. However, the Government had no expertise in the field of coastal erosion and in most cases coastal protection measures were left up to the individual property owner.

In 1983, the Inter-American Development Bank (IDB) supported a team of consultants for a 12-month period to study the erosion problems on the south and west coasts. Concurrently, the Government formed the Coast Conservation Project Unit (CCPU) as a counterpart. On project completion, CCPU became a permanent CZM research and advisory agency engaged in environmental monitoring, coastal development control (in conjunction with the Town Planning Department), advising the public and government agencies on sea defence measures, management of beach access and small scale erosion measures such as revegetation. Environmental monitoring included beach profiles and wave and current measurements. In addition, water quality and coral reef monitoring surveys were conducted in conjunction with other agencies.

In 1991 a 3-year feasibility project was started with support from IDB. This consisted of an engineering study and an institutional and legal study, both of which were to be used in the preparation of a CZM plan. As a result CCPU was reinforced and renamed the Coastal Zone Management Unit (CZMU). A CZM plan has been prepared (west and south coasts) and legislation drafted for an expanded CZM program, including pollution control. In 1997, the program was expanded further to study CZM for the lesser developed east and north coasts.

From the Barbados experience, it appears that an integrated CZM agency can evolve gradually and incrementally. In Barbados coastal management evolved in response to a specific problem, beach erosion. This would be the case in most of the small islands of the Eastern Caribbean, where islands do not set out to establish CZM programmes; rather, they set out to solve specific coastal problems. The solution and responses can then evolve into a fuller CZM programme. This was also the case in the British Virgin Islands, where there were two major problems both of which stemmed from increasing levels of coastal development: loss of mangrove habitats and beach sand mining. Concern about these problems led to the establishment of a Conservation and Fisheries Department which became the nucleus for a CZM programme.

In the Barbados case, one important lesson learned during the first phase of the IDB project in 1983 was that CZM cannot be achieved on just one coast of a *small* island; the entire coast must be considered, although emphasis can be placed on the "problem" situation.

Another lesson to be learned from this case study is the time involved in establishing a CZM programme. It may take as long as 30 years for full realization. This time span is necessary for training of staff, for public and political understanding, and for programme development. For Barbados it will be the end of the 1990s before Barbados has a fully integrated CZM programme.

The same can be said for the British Virgin Islands. But much can be achieved on an informal basis before the legislation and the full CZM plan are in place.

It is possible to implement certain measures through existing agency programmes, particularly those of planning, ports, fisheries, and public works agencies. While coordinated multiple agency involvement is the key to successful CZM, it is clear that one government agency must take the lead responsibility for CZM. Interagency committees are a very important part of the process of integration.

Contributed by: G. Cambers, University of Puerto Rico, Sea Grant College Program.

8.5 *Bonaire: carrying capacity limits*

The coral-rich waters surrounding Bonaire in the Netherlands Antilles are a major tourist attraction and have been designated the Bonaire Marine Park (BMP). However, the steady increase in the number of divers visiting BMP (from about 50 000 in 1981 to 180 000 in 1991), caused concern about diver impacts and the sustainability of the coral based resource. The World Bank commissioned a study of the situation in 1991.

Fortunately, an extensive reef mapping project conducted in 1981–82 provided good baseline data for comparison with the 1991 situation. The study focused on a comparison of two heavily dived sites and one moderately dived site. "Control" sites were selected which were similar to the test sites except that they had received little diving activity. It was assumed that if significant differences in coral cover were found between a heavily dived site and a control site they could be attributed to the impact of divers.

The percentage of live coral cover was found to be significantly lower in the heavily dived sites than the controls. Also, comparison of the 1981–82 estimates of coral cover with 1991 found that the percentage of live coral cover was significantly lower in the heavily dived sites, but not in the moderately dived site. The comparisons between sites and over time clearly indicate that there were impacts from recreational use.

The study arbitrarily concluded that the decrease in coral cover observed at the two heavily dived sample sites was not acceptable and in fact exceeded the aesthetic carrying capacity of those particular sites. The range extent of the "unacceptable" impact was more than 100 meters and less than 260 meters from the mooring buoy location. Dive statistics of the heavily and moderately dived sites showed that the impact becomes aesthetically unacceptable when a site receives consistently over 5000 dives per year. A number of popular dive sites already had more than 5000 dives per year in 1991, thus exceeding the presumed carrying capacity for individual dive sites.

The total "diveable" coastline was estimated at 52 kilometers, and with moorings spaced 600 meters apart the Park could have a total of 86 dive sites. At 4500 dives per year per site there could be 387 000 dives per year. Making corrections for unevenness of use of the various sites the maximum carrying capacity was set at about 200 000 dives per year. Concurrent with Bonaire's recommended total tourist limit of 40 000 divers (of a total of 100 000 tourists) the 200 000 dive limit would allow five dives per visitor.

An important lesson learned is that for the use of marine resources for tourism to be sustained, tourism interests must be warned early on that resource use may have to be limited at some stage. It is essential to assess carrying capacity at an early stage of development and to refine the assessment later.

Source: Adapted from T. van't Hof in *Coastal Zone Management Handbook* [1].

8.6 *Canada: offshore sewage outfall at Victoria, BC*

A marine offshore sewage disposal system was developed over the past 25 years by the city of Victoria, BC, on the Pacific coast of Canada, which lies alongside a well-flushed Strait with a net offshore drift. Victoria opted for a system which screens, *but does not treat*, sewage before discharging it well offshore (1.1 and 1.7 kilometers). The system efficiently handles a population of about 300 000 people.

The issue about this system (completed in 1980) is whether conventional *sewage treatment* should be added. The answer is plainly, no. In the author's view, this would waste money and create no significant benefit. But there was some local feeling in Victoria that discharge of untreated sewage to the sea is in principle environmentally undesirable, in spite of proof of insignificant impact since 1971.

The options available to any sewerage authority range from no treatment, through simple mechanical procedures such as screening, to progressively more complex treatment systems customarily labeled primary, secondary, and tertiary treatment. The effluents of such systems may still contain high (ecologically damaging) levels of nutrients and some pathogens and solids (except tertiary, which is very expensive).

All treatments reduce pathogens, suspended solids, biological oxygen demand (BOD), and bad odor. But all treatments also produce sludges, which may be infectious, loaded with toxins, and have a high BOD. These sludges must then also be disposed of in some way which is not polluting.

Screening for offshore disposal does not produce a sludge. It removes coarse solids which are disposed to a landfill.

At Victoria's two ocean outfalls (60+ meters deep), moving seawater kills pathogens, and dilutes, disperses, and recycles nutrients. The dispersed solids deposit below resuspension depths. The design criteria for both was essentially that the sewage plume could rise only to, and disperse at, a trapping depth below a sharp density gradient (the "pycnocline"). In addition, the dispersing layered field could not drift to shore.

A technical team found through review of monitoring data that moderate chemical contamination occurs to a range extent of 400 meters and at lesser levels out to 1600 meters maximum distance from the outfalls. Biological effect was even more restricted. Moreover, there was no detectable effect on the shoreline.

In summary, the Victoria system has been successful in collecting and disposing of human wastes and preventing epidemics of water-borne infectious diseases while producing very limited environmental change.

The following were the main lessons: (i) most important was to learn that a sewage disposal system for a coastal city may be able to achieve public health protection with low environmental impact for minimum cost by discharging screened sewage offshore; (ii) it is essential to undertake adequate Environmental Assessment (EA) prior to design of such systems and to continue monitoring indefinitely in order to ensure that the design criteria are being maintained; (iii) expeditious public release of information about the system is essential; and (iv) enough funds to maintain the EA program over time must be made available.

Source: Adapted from D.V. Ellis in *Coastal Zone Management Handbook* [1].

8.7 Ecuador: situation management

Ecuador's coast has fostered a large and growing shrimp mariculture industry which has transformed most of Ecuador's estuaries into shrimp ponds to make Ecuador the world's largest producer of farmed shrimp as of the early 1990s: 133 336 hectares of ponds generated some US$526 million in exports.

The shrimp farmers have created major problems, including constricted water flow in estuaries, destruction of mangroves, increased use of agrochemicals, and flows of untreated sewage. The ensuing degraded water quality has even affected the productivity of shrimp farms that cause it. Along most of the 2859 kilometers coastline there is no effective governance structure for resolving the mounting conflicts among user groups or considering the consequences of development.

The project, begun in 1986, adopted a two-track strategy of building constituencies and governance structures simultaneously at central government (Track 1) and community levels (Track 2). The Programa de Manejo de Recursos Costeros (PMRC) was formally enacted by Executive Decree in 1989.

The PMRC has been built on the principle of decentralization; emphasizing response to needs for both development and conservation and involving stakeholders in each step of the management process. The program began with a highly participatory process of issue identification and analysis of the subject coastal provinces.

In planning, the program focused upon developing detailed CZM Situation Management plans for the five Special Area Management (SAM) zones. Each plan addresses five generic issues: shrimp farm development and fisheries, land use in the immediate watershed, use of the shorefront, environmental sanitation, and mangrove management. The planning process has featured "practical exercises in integrated management", such as latrine building and beach cleanup.

A major lesson learned is that local capacity in resource management should be built simultaneously at both the community level and within central government. The biggest challenge is to build constituencies for improved resource management and to design governance systems adapted to the socio-political traditions of the country. In a country which has seen little success in implementing broad resource management initiatives, a focus on geo-specific forms of Situation Management like SAM is an effective strategy.

Source: Adapted from S. Olsen in *Coastal Zone Management Handbook* [1].

8.8 Egypt, Sinai: planning for biodiversity protection

With an area of 67 000 hectares, Lake Bardawil is the largest lagoon in Egypt. Situated in the north of the Sinai Peninsula it is about 85 kilometers long. The hyper-saline lake is influenced by an arid desert climate (<100 millimeters of rain per annum). Because of the arid conditions, the shorelands have long been underdeveloped and sparsely inhabited by Bedouins with a traditional nomadic life-style that has little impact on the desert environment.

The lake is an important area for migratory waterfowl, acting as a wintering site for significant numbers of Palaearctic birds, including white pelican, garganey, and little stint. The lake has been put on the list of wetlands of international importance of an international wildlife treaty (the Ramsar Convention). With other Egyptian lakes becoming progressively more polluted, Bardawil will increasingly become a more important wetland refuge for migratory birds.

Besides its biodiversity values the lake provides significant breeding grounds for Mediterranean fish species. The lagoon also supports an important fisheries industry with yields reaching some 2500 mt of fish annually, with export values of over US$12 million. Production and species composition depend largely on maintaining an optimal hypersalinity level (from 45 to 55 ppt).

In addition to the lagoon's fisheries and biodiversity significance, the lagoon lies in a region with a profusion of archaeological sites. Over the last 5 millennia, the north coast of Sinai has served as a "bridge" between

Eurasian cultures and civilizations of the Nile. The coastal stretch is littered with hundreds of yet unrecorded monuments.

In the absence of an institutionalized mechanism for CZM, numerous developments underway or planned will lead to substantial changes of the ecological and socio-economic conditions around the Lake. Tourism development started in the early 1980s when star-rated hotels and resorts were built around the capital town of El Arish, less than 25 kilometers from the Lake's east end.

Tourism development continues in a westward direction toward the Nile Valley (and Cairo) along the main transportation artery which lies just south of the lakeshore. A development with a far more significant potential impact is the proposed construction of the El-Salam irrigation canal which seeks to divert freshwater from the Nile, through a canal beneath the Suez canal to feed into a 170 000 hectare large irrigation scheme, Northern Sinai Agricultural Development Project (NSADP). The proposed project would be located along the Mediterranean coast, engulfing Lake Bardawil. As part of the Project, some 22 000 families from among the rural communities of the overpopulated Nile Valley will be given land holdings and shelter in this area. The NSADP project, if not constrained, could strongly impact Lake Bardawil and have negative ramifications for socio-economic conditions of the Bedouins.

The main impacts are: (i) NSADP would cause an inflow of fresh water to the brackish lake, lowering salinity and increasing the levels of nutrients and pollutants; (ii) new settlements would increase human waste discharges to the lake leading to eutrophication of the lagoon; (iii) settlers would likely engage in (illegal) bird harvesting; and (iv) displacement, loss of traditional land rights, and loss of cultural heritage of Bedouins.

Feasible and cost-effective measures for reducing impacts of the NSADP to acceptable levels include: (i) reducing groundwater seepage toward the lake, by matching actual irrigation supplies with actual crop water requirements to minimize drainage surplus in the sandy soils; and (ii) putting a functional CZM planning structure in place to control the intrinsic aspirations of various governmental and private entities for future exploitation of the area.

Community-based coastal management systems have a long history in Sinai and have thus far proven to be effective in managing coastal resources as demonstrated by Bardawil's rich fisheries. The Bedouins customary and legal system which permitted resource allocation among the Bedouin users is in danger of being overwhelmed by a ground swell of externally financed projects.

In response, local authorities have called for an all-embracing spatial planning regulation for the Northern Sinai and have commissioned a land use Master Plan as well. The Master Plan will allow clear designation of land to be set aside for preservation of wetland resources and antiquities, shoreline protection for beach recreation, and areas suitable for residential, commercial, industrial, agricultural, and fisheries developments. Also, the correct level of development intensity can be identified and allocated in coastal development areas. Any plans not conforming to the Master Plan would be vetoed.

Source: W.J.M. Verheugt, Euroconsult, Arnhem, the Netherlands.

8.9 *Indonesia, Sulawesi: disposal of harbor silt*

The rehabilitation and expansion of Ujung Pandang Harbor on the Straights of Makassar in Sulawesi, Indonesia, required that 1 000 000 m³ of bottom material be excavated from the harbor. Special care was necessary to prevent sediment pollution from the marine excavation operation. Standard countermeasures were required (sediment traps, silt curtains) as well as appropriate construction equipment (e.g. suction dredge rather than bucket or drag line). But the most difficult part was choosing a silt disposal method.

Two potential countermeasures were available to reduce environmental impacts from disposal of silt: (i) land disposal, whereby the material is usually placed in a contained site (diked) on land and allowed to de-water, after which it may be used for soil supplement or left as landfill; and (ii) sea disposal, whereby, non-contaminated material is barged or piped to a place where it can be dumped with the least environmental impact, and contaminated material is confined in special land or underwater containments.

In this case, land disposal and/or possible commercial use of the silt were deemed not feasible. This left only dispersed or contained sea disposal as a viable option. Containment was eliminated because we had no evidence that the silt was contaminated.

The objective of open sea dispersal is to spread the material in as efficient a manner as possible so that dilution occurs rapidly as the material is carried away with water currents, thereby rendering it as harmless as possible as quickly as practicable. It is best if the method simulates the natural process by which the sediment was originally eroded from the land and carried by rivers to the sea. Of course, some level of impact will occur even though the best dispersal methodology is used.

It was clear from published sources that the main current in nearby Makassar Straight is southerly so the silt disposed at sea would move generally south in suspension. An exception is that the current goes north, close inshore, for a time in the wet season (West Monsoon) mainly from November to December. Our limited study with a current drogue (November) confirmed this northwards flow.

To test the idea theoretically we used a scenario of dispersal of material with a median size of 15 μm (0.015), a typical silt from the Harbor bottom. We calculated 1 day of suspension for particles of 40 μm and 4 days for those of 10 µm with current transport south at a net non-tidal drift rate of 0.05 m/sec (about 0.1 knot) or 4.3 km/day (2.2 nautical miles).

According to this scenario, a maximum 60% of the silt (600 000 m³) would have been deposited on the bottom over an area of approximately 90 km² to an average thickness of 1 centimeter. At this point the suspended material could be calculated to be diluted to nearly natural background levels in 4 days.

Inside an approved "mixing zone" (a 3.5 km² rectangular disposal site) there would be localized ecological disruption (light blockage, silt fallout on inhabited sea bottom, repulsion of schooling fish, lessening of phytoplankton blooms) but this is considered acceptable because a small area would be involved and most effects would not be permanent. Starting with 150 mg/l at the boundary of the mixing zone by the end of 1 day of current drift the suspended solids should have dropped to less than 80 mg/l. By the second day the level should be down to between 50 mg/l and 25 mg/l or less, as natural dispersal and fallout processes continue.

According to this scenario, within about 4 days the silt discharged would become diluted to 17 mg/l, which added to natural background amounts (35 mg/l) would be less than that indicated in the Indonesia official guideline for suspended solids: <80 mg/l permissible for fishery and marine park which is equivalent to 7–10 NTU.

The discharge site was located where: (i) no coral reefs would be in the path of silt drift in the recommended season of disposal (April–October); (ii) where fishing pressure was light; and (iii) where biodiversity was low. Principal commercial shipping lanes were avoided.

An electronic Hach Turbidity Meter (simple, inexpensive) was recommended for oversight monitoring. But this instrument reads out in NTU units, not mg/l of suspended solids; therefore a conversion scale is needed. In this case, we calibrated the Turbidity Meter against standard solutions prepared from silt taken from the harbor bottom in the particular place of dredging (see chapter on Transparency/Turbidity Measurement in [1]).

Monitoring reports from the implemented project (1995/6) indicate that the theoretical construct held and the silt was dispersed, except that it sank a bit faster than anticipated. An unexpected result was that the silt plume attracted fish schools which were harvested by alert fishermen.

Lessons learned were: (i) with knowledge of oceanography, species and habitat distribution, and particle size, it should be quite possible to arrange for ecologically safe offshore disposal of non-contaminated silt excavated from harbors if open water with good current is nearby; (ii) silt of small particle size (say <25 µm) has little resource value for landfill, soil augmentation, or other products.

Source: J.R. Clark, adapted from Ujung Pandang Port Urgent Rehabilitation Project: Environmental Report [51].

8.10 *Maldives: an informal approach to CZM*

The Republic of Maldives uses a traditional framework for dealing with coastal issues, one based on a semi-formal and consensus driven approach. This framework has evolved over time because of the Maldives intimate dependency on the marine environment. The framework is now supported by an Environmental Commission with representatives of the various Ministries and Government Departments dealing with coastal and marine issues. The Commission identifies coastal policy needs and is supported by a Secretariat provided by the Ministry for the environment.

In the Maldives the traditional framework is built on experience gained over a long period of limited change and it works best when there is plenty of time to respond and reach consensus. When the framework tries to deal with rapid change it faces difficulty in distinguishing and responding appropriately to priority vs. non-priority issues. Also, tradition has generally invoked the fatalistic idea that deterioration of environment is an inevitable cost of improving living standards which often overlooks the use of countermeasures.

In the transition from traditional to modern approaches, effort often has been invested in less significant issues at the expense of dealing with more significant ones. Issues that have been emphasized include transient coral bleaching events and localized crown-of-thorns starfish outbreaks. Time and energy have been spent on these issues only to find that they are not as important as they seemed.

Conversely the issue of coral mining for building materials which has great social and environmental significance still has to be resolved. Action on this issue has been reactive rather than preventative; that is, concentrated on areas that have already been mined rather than on those that are at risk from future mining.

Since the traditional approach does not have experience of rapid change, the response to its shortcomings has been to impose *novel*, contemporary, ones. These novel mechanisms have, at best, been ignored. At worst they have led to conflict, confusion, and long-term rejection of the practicality of CZM initiatives. Examples of novel approaches are: wetland treatment of sewage, licensing and zoning system for coral and sand mining, and setback and elevated structure requirements.

The present Maldives solution is a compromise based on the creation of an educated cadre of locals with the range of experience to balance tradition with innovation to meet the challenges posed by rapid change. The Maldives government, being sufficiently forward thinking to encourage this strategy, has expended considerable effort in sending nationals overseas for training in the range of disciplines needed for effective CZM. The cadre of local experts is to bridge the gap between traditional and innovative management in a constructive way.

The key to identifying management mechanisms in the Maldives is not a matter of deciding whether to use traditional or novel mechanisms. It is rather a matter of deciding which mix of tradition and innovation will achieve the desired objectives within the most relevant time scale.

In small island states the introduction of a novel CZM program may not be justified (see also Section 8.17). One reason is that, in a small country with concentrated Government, integration may be achieved *de facto*, at least to the extent that existing leaders want it, because they are in nearly continuous communication.

Source: A.R. Dawson Shepherd, *Hunting Aquatic Resources*, York, UK.

8.11 *Montserrat: controlling beach sand mining*

Montserrat, an island of 102.3 square kilometers in the Caribbean, has suffered serious beach erosion problems caused in large part by mining of sand for construction materials (aggregate, plaster, and fill), which reached non-sustainable levels in the 1960s at many leeward beaches. Various controls were attempted without much success. Finally, in 1991, the government responded by closing all but one beach (Trant's). But mining continued illegally and new solutions were sought which could save the beaches and still not overly restrict development.

Two initiatives were pursued, starting in 1990: a beach monitoring programme and an exploratory education effort. The result of the latter was formation of clearly defined initiatives. The short term solution was to totally prohibit sand mining and alternatively encourage use of quarry "crusher dust" and imported sand. For the longer term, a search was commenced to find alternative sources of fine aggregate. But illegal beach mining continued, largely because builders found that beach sand was cheaper and better for plastering work.

In 1993 the government opened a beach ("the Farms", on the windward side) to mining under a permit and fee system (EC$10/0.7646m^3) enforced by beach wardens. Mining of other beaches then had essentially ceased. The damaged beaches were rebuilding naturally. Research on alternative materials continued. It seemed like the damage was now controlled. But things changed after July, 1995, because of the volcanic eruptions.

The area surrounding the approved beach was subsumed into an "exclusion zone" where all activity and occupation was prohibited because of present earthquake danger. This was followed later by resettlement of exclusion zones and a concurrent construction boom with a high demand for sand putting the beach protection programme into jeopardy. The government reacted by permitting increased importation of sand and by allowing limited extraction from another beach (Carr's) for the small builder.

Several lessons were learned: (i) the permit and fee system found favor with builders and the demand remained high; (ii) alternative aggregate is available but no one is able to successfully fill the demand for fine aggregate (for plaster); (iii) education programmes were an essential backdrop to support of the strong legislation and enforcement mechanisms; and (iv) government needs to be flexible and to encourage rapport between key players from both public and private sectors.

Contributed by: A. Gunne-Jones, Chief Physical Planner, Montserrat, W.I.

8.12 *Oman: coordination as the key*

The success of any plan that implicates different ministries, as the Coastal Zone Management Plan (CZMP) for Oman does, depends on two principal factors: (i) information exchange and cross-sectoral review; and (ii) effective coordination for plan implementation.

In the Oman plan, a lead agency assumes primary responsibility for administering these two tasks. Since the Ministry of Commerce and Industry commissioned the CZMP, this responsibility would come to rest primarily with them. However, certain specialized tasks relating to the management of the various resources fall directly into the mandates of other ministries, including: housing, tourism, fisheries, environment and water resources, communication, municipalities.

In the past 7 years the CZMP plan has been utilized to address several matters: resource reserves, beach erosion, coral reef management, oil spills, setbacks, data bases, turtle protection, and litter cleanups.

The main lesson learned is that a networking plan for CZM is feasible, whereby administration of the program is shared among the various concerned agencies.

Source: Adapted from R. Salm in *Coastal Zone Management Handbook* [1], with 1997 update.

8.13 *Philippines: success with community based management*

Fisheries are an important sector of the economy of the Philippines which is ranked 13th in world fish production. Between 10 and 15% of the total yield is taken from coral reefs. The 27 000 km² of coral reef resource is pristine in a few remote places but seriously degraded or destroyed in numerous other places, often those where there is intense fishing.

Destruction of coral reef habitats, overfishing, and a consequent decline in fish catches plague small-scale fishermen throughout the Philippines. The three island communities discussed below—Apo, Balicasag, and Pamilacan—were all suffering from deterioration of their marine environment in 1985. Destructive fishing methods in common use were explosives, fine mesh nets, scare-in techniques, and poison. Increasing poverty was forcing people to use more efficient, but destructive, fishing methods.

This motivated the Philippines to experiment with various forms of coastal management. One experiment which has proved effective for coral reefs surrounding small islands is creation of marine reserve and sanctuary combinations which encourage local community responsibility for fishery and coral reef resources. This reserve approach includes some protection for the coral reef and fishery surrounding the entire island but complete protection from exploitation for a sanctuary covering up to 20% of the coral reef area. The sanctuary was to provide an undisturbed place for fish to feed, grow, and reproduce.

This approach was applied to the three island communities in 2-year community based projects beginning in 1985. The results were increased or stable fish yields from the coral reefs. At present, the management regimes are supported by the community and are functioning without significant intervention. Implementation at the three island project sites included the five types of activities described below:

1   *Integration into the community*. During a 3-month initial period, field workers located in the community introduced the project, met with community leaders, attended community meetings, and generally became acculturated to the island situation. Baseline data were collected for later evaluation. Also pursued were: socio-economic/demographic surveys; pretest of environmental and resource knowledge and perceived problems of local people; and a survey to document the status of the coral reefs by means of substrate cover, species diversity and abundance, and several other indicators.

2   *Education*. Education was continuous throughout the project but emphasized in the initial stages. Most forms of education were non-formal, in small groups and by one-on-one contact. Focus was on marine ecology and resource management rationale and methods. During the education process, community problems and potential solutions emerged.

3   *Core group building*. It was clear that the correct way to implement management solutions was through community work groups with close ties to the traditional island political structure. Because funds were available for a community education center adjacent to the sanctuary, the first group activated was the one responsible for Center construction. Secondly, individuals interested in the conservation program formed a Marine Management Committee (MMC) to study problems of each island. Each MMC earned community respect once it decided to implement a marine reserve.

4   *Formalizing and strengthening organizations*. Other initiatives were aimed at providing continuing support, in real and symbolic terms, to the MMC such as helping it to: (i) identify enhancement projects such as reforestation; (ii) place giant clams in the sanctuary for mariculture; (iii) refine the marine reserve guidelines; (iv) train MMC members for guiding tourists to

95

the island; (v) collect fees for visits to the sanctuary; and (vi) try alternative income schemes such as mat weaving.

As a result, Apo Island became a training site whereby the MMC helps to conduct workshops by sharing their experiences from the Apo success with other fishermen groups. This activity has truly strengthened the core group and solidified support for the marine reserve among the community.

The three island-wide marine reserves that were created receive municipal administrative support. Municipal ordinances, tailored by the communities to suit their particular needs, are posted in the local language. Enforcement varies from island to island, with mostly moral support from the Philippine police.

The reserves are, with some local variations, well marked by buoys and signs and managed by island-resident committees which patrol for rule infractions by local residents or outsiders.

Community Centers function and serve as meeting places for the MMCs and other groups. Diving tourism to both Apo and Balicasag Island has increased markedly in response to the sanctuaries which are teeming with fish.

Fishermen members of the three island MMCs confirmed in 1992 and 1997 that the marine reserves and sanctuaries had significantly improved fishing by serving as *semilyahan* (breeding places) for fish. Fish yield studies in the reserves outside the sanctuaries indicated that yields have been at least stable and probably increased.

Comparison of baseline data of 1985 and 1986 with a survey made in 1992 showed an increase in fish diversity and abundances within the fish sanctuary at Apo Island (see Table 2).

Concurrently, the coral reef cover in the sanctuary and non-sanctuary areas of the three islands has remained stable and perhaps improved slightly since 1984, which is generally not the case for coral reefs of the Philippines.

Specific lessons inferred from this experience in small-scale coastal management can be of use in CZM-type Situation Management schemes in other settings, including larger coral tracts, as listed below:

1 It is possible to effectively manage small island coral reef resources as a community based, co-management effort. Benefits can be derived from a management regime which prevents destructive uses, limits fishing effort

Table 2 Comparison of fish diversity and abundance at Apo Island. (Sources: White [52]; White and Calumpong [53].)

	1986	1992	Percentage increase
Species richness	52.4	56.0	6.8
Abundance			
Food fishes	1286	2352	83
Total fishes	3895	5153	32

by establishing a marine reserve with a core sanctuary, monitors the impact of the program, and shows the real benefits and feeds results back to the resource users.

2   Small islands provide a geographical advantage to marine resource management because of decreased access to outsiders. In addition, island communities more easily identify with their own resources, as the coral reef vicinity becomes a territory over which it has some control.

3   Local residents must understand how a management program will solve a problem they think is important. For example, if they see no links between a degraded coral reef and decreased fish catches, they will not support the management action. The importance of this is emphasized by the peoples' initial belief that corals were just stones!

4   Formation of capable and respected community groups is critical for successful implementation of community resource management projects. This means more than simply a mayor's approval. It requires groups working together on projects with real outputs.

5   People must observe some immediate results if they are to continue to support management with restrictions. Education can provide the initial understanding of why a program is needed, but only observable results can sustain a program.

6   Community-based approaches which mobilize the people who use the resources daily are necessary to insure wide participation and potentially long-lasting results. Regulatory approaches have had few successes in the Philippines.

7   Complete and practical baseline field surveys and follow-up monitoring are prerequisite to helping a community design a feasible management plan and comprehend results. Data on the increase in fish abundance and diversity can be used to convince policy makers and government officials about the effectiveness of the marine reserve management.

8   Fishing communities can themselves correctly choose the size and location of a marine sanctuary with the assistance of community organizers and technical inputs from marine scientists in a mutually supportive and open environment for discussion and negotiation.

9   CZM projects need to consider linkages among all potential participants, including community leaders, municipal government, law enforcement officers, private business, and national government organizations like tourist authorities or fishery agencies.

10   Successful management for small-scale settings is vulnerable to changes in political leadership. Results are also sometimes dependent on continuing moral and physical support from outside entities not obvious during the initial implementation phase.

Acknowledgments: This work was made possible by Silliman University and staff, particularly A. Alcala and N. Calumpong. Support was provided by the East–West Center in 1983, Earthwatch International in 1992, and USAID in 1997.

Source: A.T. White, Coastal Resource Management Project, Cebu, Philippines.

8.14 *Solomon Islands: social chaos from tourism*

Anuha is a small island of 60.7 hectares lying just 0.8 kilometers off the north coast of central Ngela, about 83.3 kilometers from Honiaira, the capital of the Solomon Islands. With four hills clothed in virgin tropical rain forests, 8 kilometers of white sandy beaches, a fringing lagoon with two small palm crowned inlets on the fringing coral reefs, and a freshwater lake thrown in for good measure, it is the tropical island of tourists' dreams.

An Australian resort company negotiated a 75-year lease of the island from its customary owners, a small village of 60 people on the Ngela mainland, led by a retired Anglican priest, Father Pule. There were claims of ownership of the island by other Soloman Islands groups who 150 years earlier were granted refuge on Anuha Island by Ngela, as they sought escape from the marauding head hunters of Marovo. But the claims failed in court.

This was the first resort of any size in the Solomon Islands. Therefore, the lease agreements were seen as a test case for foreign investments and a model for future tourism development. The original Australian shareholders had been sympathetic to the customary owners; for example, taking care to pay compensation when they had to remove a tree of special value to the owners. They made a point of employing local villagers.

But the second owners cared little for *kastom*, the traditional way. They considered that the lease gave them the right to do with the island as they saw fit. First they dismissed the original expatriate management team. Then they dismissed many of the local workforce, replacing them with tribesmen from other islands. They bulldozed a section of the rain-forest to double the number of bungalows without consultation or compensation, and insulted and demeaned the customary owners.

Father Pule was incensed. He claimed there was no agreement with the new owners and the lease agreement was void. He began a campaign for a new lease, seeking a single US$2 million lump sum payment while future generations were forgotten in pursuit of immediate wealth. Consequently, he had his followers invade the island and dig holes in the airstrip. Other incidents included a Christmastime raid when Father Pule sent warriors in war paint to force guests off the island, close down operations, threaten non-Ngela employees (coercing them to flee the island), and damage equipment such as outboard motors.

The ownership then passed to a third investor (August 1987) who was unaware of the legacy of distrust he inherited. When an Australian chef fell out with the new management and turned to Father Pule, the stance of the customary landowners changed from demanding a new lease to demanding compensation and repossession of their island.

Father Pule's warriors again invaded Anuha (apparently at the exhortation of Smith), forced the expatriate management team off the island at spear and arrow point, and held guests and construction crew hostage for several days. Some parts of the resort were damaged. Finally the Royal Solomon Islands Police field force went to the rescue and detained a number of Father Pule's men.

The chef, John Smith, gained access to Government Ministers, including the Prime Minister. Death threats against the Australian High Commissioner and his staff by Smith (who alleged to have his own band of warriors) resulted in protection of diplomatic residences by the Solomon Islands Police.

Next (May, 1988) the resort's central complex burned down with a loss of about US$1 million. Smith was suspected of arson and received a prohibition order (1988). The resort remained closed. Father Pule refused to negotiate further and a court order was entered against him and his followers for damages and wrongful entry. They were forbidden to visit the island unless employed by the resort.

Then Anuha's owners wanted to back off from their case to resolve the deadlock, but Father Pule refused to re-enter negotiations. In mid-1989 Father Pule's men replanted the airstrip with coconut trees. The trees were removed but the resort remained closed (May, 1990), failing as a tourist destination. No tourist operator can stay in business for very long if he must ring his resort with barbed wire fences and employ armed guards to keep tourists in and local people out.

This conflict had far-reaching consequences on investor confidence in the Solomon Islands, on the security of foreign businesses and foreign residents, and on tourism to the country. It tested the legal system and found it wanting in fair and constructive management of disputes involving *kastom*, by favoring expatriates. For his efforts, Father Pule benefitted by clear and unambiguous ownership of the island by the villagers from Ngela.

Source: Adapted from T.H.B. Sofield, James Cook University, Townsville, Queensland, Australia [54].

8.15 *Sri Lanka: issue-based and incremental CZM*

As the first tropical country with a centrally managed, full-scale CZM program, the Sri Lanka case demonstrates how coastal zone program creation is often issue driven. The original issue was coral mining and consequent jeopardy to the shoreline.

The removal of coral materials, stimulated by a surge of economic development in the 1960s and 1970s, was so extensive that it left the shoreline increasingly exposed to natural hazards. Serious erosion of beaches led to loss of shorelands and exposure of the coast to storm surges. Also, mangroves, small lagoons, and coconut groves were lost to shore erosion and local wells became contaminated with salt water. Coral mining also contributed to collapse of a local fishery.

In 1983, Sri Lanka responded by enacting a national CZMP under administration of the Coast Conservation Department (CCD). Jurisdiction extended 300 meters landward and 1 kilometer seaward of the coastline and included estuaries and river shorelines for 2 kilometers inland. The program was successful and in the 1990s was expanded to also include protection of Special Habitats along with scenic and cultural/religious sites and to decentralize the program by sharing authority with communities.

The CZMP, which was developed in close coordination with other Sri Lankan agencies, provides both a policy framework and a practical strategy for dealing with coastal issues. CZMP strategies include regulation, research programs, enhanced intergovernmental coordination, and public education. For example, the erosion management strategy establishes a setback line to ensure that structures are not sited too close to the shoreline. These management efforts are integrated with a public works program for appropriate shoreline protection works.

While there is a general setback of 60 meters wide, the demands of development do not always permit reserving a setback of this width, so there are variations according to physical features of the coast and some exceptions. Setbacks for particular *water-dependent* activities such as hatcheries for aquaculture, boatyards, etc., are determined on a case-by-case basis in the Sri Lanka program. In addition, "performance standards" have been utilized which emphasize environmental results from a project rather than impose specific construction rules, freeing developers to innovate.

Other regulatory measures prohibit the construction of shoreline protection works in some locations and set review procedures for the rest of the coast. In addition, coral and sand mining are controlled. Also included in the erosion management strategy are an education campaign for coral and sand miners, a program to identify alternative employment for displaced coral miners, and identification of alternate sources of lime for the building industry.

Attached to the CZM process is a policy group, the Coast Conservation Advisory Council (CCAC), which reviews all major permit applications along with the accompanying Environmental Impact Assessment (EIA) and CCD staff recommendations. In virtually all cases the CCAC agree with the staff and support their decision and so advise the Director of CCD. This is a very important aspect of the Sri Lanka CCD and seems to operate with a minimum of political interference.

CCD has begun the process of delegating authority for issuing "minor" permits to local authorities and has organized an extensive program of training local officials. This "devolution" process also requires a local monitoring system for insuring that permit procedures comply with the Coast Conservation Act. A major criterion of success for this effort is how well sharing authority between national and local levels of government works. Local involvement is the key to finally putting an end to coral mining.

The "bottom-up" approach of decentralization by CCD also involves the use of Situation Management, specifically the designation of special area management (SAM) sites in which residents become actively involved in both the design and implementation of the coastal management program. SAM projects rely primarily on reciprocation through mutual education, persuasion, and cooperation to minimize activities that deplete or degrade coastal habitats to the detriment of all residents. Trial SAM projects are underway in two south-west coast sites, Hikkaduwa and Rekawa Lagoon.

Lessons learned are: (i) it is feasible to start with a single issue and then expand the CZM program; (ii) alternate employment is required for displaced workers; (iii) active participation of stakeholders is necessary; (iv) sharing authority with communities is beneficial; and (v) use of performance standards encourages project sponsors to innovate environmentally safer designs and methods.

Source: Adapted from Olsen [55], CCD [56] and Clark [1].

8.16 *Sri Lanka: Negombo Lagoon zoning*

With an area of some 7000 hectares, the Muthurajawela Marsh–Negombo Lagoon (MM-NL) ecosystem lies on the west coast of Sri Lanka, 15 kilometers north of its capital Colombo. The lagoon proper (3200 hectares) is connected to the Indian Ocean by means of a narrow inlet near the town of Negombo (see Fig. 12). The lagoon supports about 3000 fishing households and provides numerous environmental services with an annual worth of over US$23.5 million.

MM-NL is of international significance for biodiversity and as a refuge for migratory birds. But the Lagoon and surrounding marshland area are beset by a profusion of socio-economic and environmental problems. They include: (i) prevailing poverty of fisherfolk causing overexploitation of the marine and brackish fisheries resource; (ii) changing land uses in the catchment area which have altered runoff and sedimentation patterns and changed key hydrological characteristics of the lagoon; (iii) rapid population increase in the north Colombo suburbs which leads to illegal encroachment and puts pressure on the wetlands fringing land area; and (iv) land use decisions not based on the complex ecological systems of MM-NL.

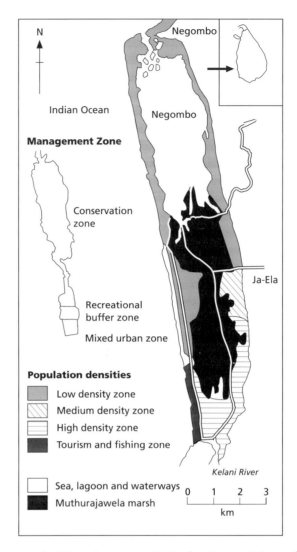

Fig. 12 Zoning plan for Negombo Lagoon, Sri Lanka. (Source: J. Samarakoon, H. van Zon and W.J.M. Verheugt.)

An ecosystem-based approach was used to integrate environmental considerations into an MM-NL Situation Management type Master Plan. The plan was prepared by a consultant team based on biological, geo-physical and socio-economic resource information and consensus building among stakeholders. Support was provided by the Netherlands government during a period of over 7 years.

The foundation of the Master Plan was zoning, which addressed the issues of development needs, conservation importance and equity. Detailed mapping at 1 : 10 000 scale was undertaken to facilitate zoning. For planning purposes, four zones were delineated for the lagoon and surrounding area (a total of 10 700 hectares), endorsed by stakeholder workshops. The workshops were: (i) Conservation Zone (91% of wetland); (ii) Buffer Zone (6.4% of wetland); (iii) Mixed Urban Zone (2.5% of wetland); and (iv) Residential Zone (42% of total planning area).

The Master Plan received Cabinet approval in 1991 and a Master Plan Implementation Steering Committee (MPISC) was established. By 1994, five main types of activities were identified for further planning: (i) relo-cation and community development package for 200 households who had encroached upon the Conservation Zone; (ii) an EA requirement for proposed developments in the Mixed Urban Zone and a detailed land use and market-ing plan; (iii) a management plan for the Conservation Zone; (iv) a land use plan (after community consultation) and screening of investment proposals for the Buffer Zone, aiming at economically viable tourism development; and (v) development of a cost-recovery system for conservation management.

The MPISC was instrumental in ensuring community participation during the entire planning process. The workshops resulted in a common vision on the ways to tackle the main coastal zone management issues, obstruction of lagoon-water exchange due to heavy siltation, and destruc-tion of fish nurturing areas. It was agreed that basic objectives were sustain-able use of the lagoon resources, community development, pollution control, enforcement of environmental legislation, and creation of job opportunities in tourism.

The major lessons that emerge from this experience are: (i) the need for practicality, legitimacy, and equity; (ii) the need for a strong scientific and technical foundation based on ecosystem structure and functioning; (iii) the need for community and stakeholder involvement and empowerment; and (iv) the need for high level political commitment and interagency coordination.

Source: Dr J. Samarakoon, H. van Zon and W.J.M. Verheugt.

8.17 *Trinidad and Tobago: choosing the non-CZM option*

Trinidad and Tobago, like most small island developing states, is faced with a variety of problems which challenge the achievement of sustainable develop-ment. Small island nations have several problems: their size, limited natural resource base, fragile ecosystems, susceptibility to disasters, depletion of non-renewable resources, economic openness and development pressures caused by increasing population demands which contribute to environmen-tal degradation.

Sustainable development often requires assistance from the international community. Technical and financial assistance are required in the areas of capacity building and institutional strengthening, waste disposal and management, integrated management of natural resources, management of the impacts of natural and anthropogenic disasters, and implementation of integrated development planning.

In small island nations, solutions to coastal area or environmental problems often involve the whole island, including the Coastal Zone. National planning in Trinidad and Tobago was initially just for economic development, but in 1969 the concept of regional planning was introduced into the national planning and development process.

The National Physical Development Plan was prepared by the Town and Country Planning Division (TCPD) of the Ministry of Finance and Planning and approved by Parliament in 1984. The Plan sets out to formulate a coherent and comprehensive land use policy which will provide the criteria for the execution and enforcement of development control. It seeks to ensure consistency and coordination by providing guidance on matters of land use to government agencies as well as to private developers and investors. The Plan provides a framework for the preparation of regional and local plans, and for the integration of spatial planning with socio-economic sectoral policy making. The Plan articulated the need for an environmental management policy and stated the strategy for environmental planning and conservation.

The idea that there was a need to establish a separate agency to deal with management of the Coastal Zone in Trinidad and Tobago was introduced as early as 1974. Ten years later TCPD provided opportunity for inclusion of the land sector (only) of the coastal zone. But there were barriers to success: (i) the lack of knowledge and shortage of expertise; (ii) the institutional problem; and (iii) the jurisdiction problem.

The proposed solution to problems (i) and (ii) was the establishment of a special multi-disciplinary coastal agency. But such an agency would not, of itself, solve existing jurisdictional problems. It was concluded that a full-scale coastal area development and management study should be undertaken using the West Coastal Area of Trinidad as the pilot area.

The Institute of Marine Affairs (IMA) undertook a multi-disciplinary coastal area planning and management study. The intent of this study was to suggest a pattern of growth which would permit and encourage development in a balanced manner. The West Coastal Area of Trinidad was to be used as the priority coastal area. The techniques developed in the study would be utilized for a nationwide Coastal Area Plan.

At the end of 1984, it was decided that the IMA should integrate its efforts with those of the TCPD to develop a unified approach to CZM. IMA could contribute to three main areas of TCPD's work: development plans, research projects, and review of applications from developers. Planning aspects would be covered by TCPD while IMA would deal with environmental aspects. IMA continued to carry out EAs involving land and water uses for the coastal area development study as envisaged. Finally, the idea of a special CZM agency was abandoned when IMA recognized the National Physical Development Plan as including the coasts.

The lesson learned from Trinidad and Tobago's experience is that in small island countries with limited resources, there may be insufficient motivation to establish a separate CZM authority. An alternative is to have the function handled through existing planning, research, and administrative machinery.

Source: H. McShine, adapted from *Coastal Zone Management Handbook* [1].

8.18 *United States: a dollar-based program*

Operating for over two decades, the CZM program of the United States has used a combination of grants and organizational incentives to enlist 29 states and territories covering 94% of the nation's shoreline in an effective intergovernmental coastal management network. At a relatively low cost to the federal treasury, the program has resulted in major improvements in land use regulation, natural resource protection, public access, urban waterfront redevelopment, hazard mitigation, resource development, and ports and marinas.

Key goals stated in the Coastal Zone Management Act (CZMA) of 1972, are to manage growth so as to *balance* conservation and development under a *voluntary* national program with guidelines that allow individual states to design their coastal plans to meet their particular needs and that require federal activities to be *consistent* with the federally approved state plans.

Features of CZMA are: (i) the use of financial and other incentives to create a coastal policy coalition spanning national, state, and local governments; (ii) the ability to deal with the varied geography and management problems of the US coast; and (iii) the dynamics of a participatory democratic approach, in which attitudes of any new national administration cannot easily dismantle the established CZM program.

In forming its CZM, the US Government considered two program designs: (i) a centralized national program; and (ii) a decentralized, financial incentive-based program where states participate voluntarily, under broad national guidelines.

The solution chosen in 1972 was a mixed alternative. It incorporated both bottom-up state plans and top-down oversight by the national government. It required balance between development and conservation goals within a voluntary, decentralized program in which individual states and territories were strongly encouraged to participate through federal financial grants for planning and implementation. Each participating state must submit its plan to the national coastal agency for approval, and in return is promised that future federal actions in coastal areas will be consistent with approved state plans. The bulk of federal dollars has been spent by states on two types of activities: improving government decision making and protecting natural resources.

The technical basis of the program focuses on comprehensive land use planning and environmental controls as the means to protect coastal resources. It encourages states and localities to use a variety of innovations. Much of the success of the program is owing to an extensive base of environmental regulations administered by a variety of state and federal agencies. CZM coordinates these regulatory programs toward unified management.

A lesson learned is that a unified CZM program like that of the US CZM program is durable and effective because of the intergovernmental collaboration that springs from federal funding and oversight, voluntary state participation, and the promise of federal consistency with approved state plans. States enter the program because they want to, rather than because they are required to. This recognition of the states as responsible co-managers builds their commitment and loyalty. Rather than federal control, the spirit of the US dollar-based coastal program is based on collaboration and capacity building.

Source: Adapted from D.R. Godschalk in *Coastal Zone Management Handbook* [1].

8.19 *United States, Hawaii: the ocean outfall solution*

Kaneohe Bay, along the north-east coast of Oahu, is the largest embayment in Hawaii, measuring 11.2 kilometers long and 3.2 kilometers wide. A broad barrier reef extends offshore across the mouth of the Bay, maintaining a lagoon inshore with numerous pinnacle and patch reefs. Two passes, one in the north and one in the south, bisect the barrier reef to allow tidal exchange between the inner bay (lagoon) and ocean. Fringing reefs line the entire shoreline inside the Bay. The southern third of the Bay is a nearly enclosed basin surrounded on three sides by land with restricted water exchange with the rest of the bay due to broad shallow reef flats and islands. Before 1939 the coral reefs in Kaneohe Bay were the best developed and most diverse of any in Hawaii.

Beginning in 1939, a US naval air station was constructed on the southern peninsula (Mokapu) fronting the Bay, resulting in the dredging and filling of coral reefs mostly within the southern lagoon. Over 11.47 million cubic yards of reef materials were dredged with most placed over reef flats to enlarge the land area of the naval base. Excess sediments from the dredging operations were discharged back into the lagoon. A ship's channel was also dredged along the entire length of the inner bay and the northern pass deepened to accommodate ship navigation between the Bay and ocean. Shopping centers and residential neighborhoods later sprang up, mostly in the southern end of the Bay at Kaneohe town.

By the early 1960s the need for improved sewage disposal led to the construction of a sewage treatment plant and outfalls in the Bay. The navy sewage outfall was placed at the south-east corner of the Bay while a new municipal outfall discharged into the southern basin near the Kaneohe Town. As the population grew, sewage discharges increased. By 1977 the combined effluents of three sewage treatment plants totalled more than 7.5 million gallons (20 000 m³) per day with 95% of the sewage discharged into the southern lagoon. Even though the level of sewage treatment was good (secondary), sewage discharges in the south bay were still stressing the coral reefs. Secondary treatment actually generates labile forms of nutrient (nitrogen and phosphorous) that are readily taken up by marine plants (phytoplankton and seaweeds) which in turn enables them to better compete against corals.

Reef decline began with the military dredging operations between 1939 and 1950, mostly in the southern lagoon. By the late 1960s when the first extensive surveys of coral reefs were conducted in the Bay, very little of the reef had recovered from the previous dredge and fill operations and more south bay reefs had deteriorated because of poor water quality attributed to sewage discharges. The dissolved nutrients in the effluent were stimulating massive growths of phytoplankton and eutrophic conditions which in turn led to the proliferation of suspension and filter feeders (soft corals, tunicates, bryozoans, feather-duster worms, clams, etc.). Corals and other normal reef life were all but eliminated from the south lagoon. Periodic freshwater floods and discharges of eroded soils from land contributed to the further demise of the reefs in the lagoon. The lack of oxygen and presence of toxic sulfides in the sediments may have caused some of the coral deaths in the south lagoon.

Furthermore, reefs in the middle bay also began to decline in the mid 1960s. Growths of a green bubble alga (*Dictyosphaeria*) formed massive mats, and overgrew and smothered most living corals on the reef slopes inside the Bay. Marine scientists believed that the algae were stimulated by higher nutrient levels contributed by sewage moving up from the south bay. In contrast, bubble algae coverage was much lower in the north bay where greater water flushing discouraged high nutrient levels.

Beginning about 1970, the continuing decline of Kaneohe Bay's coral reefs convinced the local government to move the discharges from the south bay to a new deep ocean outfall off Mokapu peninsula. Within months of sewage diversion from the south lagoon in 1978, water quality conditions throughout the entire lagoon improved dramatically. Within 6 years, middle bay coral reefs showed remarkable recovery while south bay reefs showed dramatic recolonization. In this initial phase of recovery, corals were rapidly recolonizing all previously degraded coral habitats. In contrast bubble algae (*Dictyosphaeria*) declined to one-fifth of the previous levels by 1983. Coral coverage also more than doubled in the Bay between 1971 and 1983 and showed increases in all areas of the Bay except in the north where live coral cover was already high.

From 1978, sewage discharges outside the Bay at the Mokapu ocean outfall have not caused problems because of rapid dilution and mixing of sewage with open waters. The outfall is also placed at more than a 30.5 meter depth to promote dilution and reduce sewage effects at the surface. It may seem enigmatic that the same amount of sewage that caused catastrophic damage to corals inside the Bay would have only negligible impact on the reefs outside the bay. The simple answer is that faster flushing time of water outside the Bay prevented a build up of nutrient levels, phytoplankton, and eutrophic waters that occurred inside the bay.

In 1990, the coral reefs in the lagoon were re-surveyed at the same sites and revealed that coral recovery had slowed down. The most dramatic increases in live coral cover occurred in the south bay. In the middle and north bay, coral coverage remained about the same (at a high level). However, the green bubble algae had begun to make a comeback in the bay, already achieving about half of its 1971 levels. This increase had not yet led to the demise of corals, but the trend, if continued, would lead to decline of corals from the smothering growths of the algae. The likely sources are cesspools, improperly maintained septic tanks, raw sewage "bypass" discharges into the Bay during power outages, sewage leaking from the aging sewerage system underneath Kaneohe town and storm water runoff. Another round of research will frame the remedial actions needed to protect the coral resource.

A lesson learned from Kaneohe is that discharge of secondary treated sewage into a poorly flushed bay is like dumping sewage into a swimming pool. Regardless of the level of sewage treatment, locating an outfall in a poorly flushed lagoon is perhaps the worst of all alternatives from the standpoint of ecosystem health, public health, and recreational/commercial values. The confined nature and slow turnover of bay waters magnify the eutrophic effects of sewage effluent. It points to the need to site sewage outfalls in well flushed waters to avoid adverse effects to coral reefs which favors an offshore outfall.

Another lesson is that careful evaluation of ecological, recreational, commercial, and other values of proposed receiving waters should enter into the designing of sewage systems along with standard engineering, sanitation, chemical, and mixing parameters.

Source: Adapted from J.E. Maragos in *Coastal Zone Management Handbook* [1].

9: The Coastal Professional

Thought involves labor; being, indeed, the hardest kind of labor, which all of us seek most to avoid. [Thomas Gibson Bowles, 1886]

In recent years the emerging field of Coastal Zone Management (CZM) has offered new professional opportunities and created new educational needs. CZM is a multi-disciplinary pursuit that involves more than 20 fields of study, from engineering to law and from chemistry to biology.

The shortage of trained CZM generalists is often cited as a reason why management cannot be initiated. In response, this chapter lists 15 universities that train CZM professionals and describes opportunities for mid-career training. The basics for a CZM professional are *first*, a degree in one of the standard technical disciplines, *second*, specific training and/or experience in coastal zone matters, and *third* a good level of know-how with computers, including electronic communication and data handling.

9.1 *The professional*

CZM is recognized as an interdisciplinary pursuit, not as a distinct discipline. Those who are its practitioners, the CZM generalists, usually have prior training in a particular subject such as, biology, law, chemistry, geology, fisheries science, hydrodynamics, and so forth.

Only recently has formal graduate level training been available to those who would make CZM a career. Those few professionals who have prepared themselves to understand the interactions of sea and land in a management perspective have been the key players in CZM work. As CZM professionals they are familiar with most subjects in this book.

9.2 *As steward*

As a CZM professional, you become an advocate of good conservation practice based on accurate technical input. Clarity and specificity of program elements are needed to convince policy makers to make a strong commitment to CZM. This effort must continue through the entire process, i.e. through all stages of development of the CZM program [57]. The following are examples of some of the initiatives you may be supporting.

1   *Reduce contamination* of the coastal zone from pollutants: industrial wastes, sewage, sediment, farmland runoff (fertilizers, pesticides, animal wastes), etc.

2 *Conserve valuable habitats* such as coastal waters, coral reefs, seagrass meadows, mangrove forests, marshlands, kelp forests, littoral embayments (estuaries, lagoons), etc.

3 *Evaluate development projects* that significantly alter water and land use in the Coastal Zone using methods such as Environmental Assessment (EA) combined with social and economic assessment.

4 *Recommend safeguards* to protect environmental resources from depletion, e.g. habitats, fisheries, water quality, and threatened species.

5 *Identify hazard protection methods* to reduce potential damage from natural hazards such as cyclones (typhoons, hurricanes), shore erosion, sea level rise, etc.

6 *Enhance multiple use* of resources including promotion of economic development that is compatible with conservation using current management techniques such as use zoning.

7 *Promote restoration* projects to rehabilitate damaged coastal environments in planning and design elements.

8 *Assist with resolution of conflicts* among the various users of the Coastal Zone and of coastal resources through issue analysis, data base preparation, mediation techniques, etc.

9 *Provide knowledge* about natural resources, the effects on them of various uses, their carrying capacities under different use loadings, and mitigative techniques.

10 *Expedite protected area designation* in the Coastal Zone for conservation of natural resources and for protection of scenic values and biological diversity.

11 *Encourage stakeholder participation* in coastal management in order to ensure extensive dialog, input, and broad support.

12 *Educate the citizenry* about coastal resources matters in order to deepen public understanding about the need for conservation and the measures necessary to achieve it.

13 *Create a strategy* for coastal resources conservation and contribute to national planning for countries that have central economic planning (e.g. 5-year plans).

In convincing supervisors, decision makers, and legislators of the essentiality of these CZM elements, nothing is more important than persuasive economic evidence. National income, foreign exchange earnings, employment, and local self-sufficiency are most important factors. These factors are addressed during the policy formulation stage of CZM and articulated during the strategic planning stage, which is the principal opportunity for selling the CZM program. Because of the importance of economic justification, CZM professionals should have introductory training in economics.

9.3 *As planner*

Regardless of your past experience, as a CZM professional you will become a planner of sorts too. The planner's role is to deal with great complexity and reduce it to simple concepts and program means that are politically and administratively viable. Typical managers, engineers, politicians, and most economic planners are not usually well informed about the sea and the sea coast at the beginning of CZM. Consequently, your special expertise will be most valuable.

CZM planning is above all multi-sectoral. In the opposite approach — narrow single sector planning — it is possible to irreversibly destroy a resource and foreclose future options for use of that resource. CZM professionals attempt to avoid this by broad multiple-sector planning and project development, by future-oriented resource analysis, and by applying the test of sustainability to each development initiative.

9.4 *As investigator*

CZM investigatory capacity is required for a variety of tasks, including the following: (i) examine ocean and coastal processes, their relationship, and impact upon the resource endowment; (ii) evaluate, develop, and conserve the resource endowment; (iii) evaluate safety and environmental protection requirements in marine areas as well as those related to the protection of life and property in coastal areas; (iv) evaluate information needs for monitoring the effects of present and planned activities on the marine and coastal environment; (v) assess new or improved technology to support coastal zone surveying, mapping, and research programs; (vi) check approaches for multiple use, and identify and try to resolve potential and current conflicts between the various users of coastal seas (see Section 6.11); (vii) investigate management structures for planning and coordinating development activities in the coastal zone; and (viii) examine regulatory frameworks and coordination mechanics needed to implement a CZM plan.

These tasks will often require the assistance of specialists because no one CZM professional is expected to have advanced training in all the disciplines such as biology, law, geology, chemistry, fisheries, and so forth.

In day-to-day operation, CZM workers often become detectives, who solve the mysteries of coastal environmental problems, such as: eutrophication of coral reefs; occurrence of red tides, ciguatera, and coliform bacteria; nitrification and salinization of groundwater; fish depletion; habitat degradation; hinterland activities that pollute the coast (agriculture and others); obliteration of wetlands; and many other problems.

9.5 *As analyst*

Much of the effort of CZM workers is involved with bridging the gap between formative policy statements and program development outcomes. The first priority is information needed for getting a CZM program approved (or disapproved, if it should be discovered that the country does not need or want CZM for whatever reasons).

All modern data handling facilities are electronic. With the advent of reliable low-cost computer systems, computer storage and analysis of geographically oriented data bases is now widely available. These computerized Geographic Information Systems (GIS) are now available on PCs (personal computers) and simple work stations, putting the equipment within the budget of most CZM institutions (see Section 6.10).

9.6 *As survey designer*

Designing field survey programs is an important job for the CZM professional (see Section 6.9) and should be done in collaboration with staff or consulting scientists who will do the studies. The survey, mapping, and evaluation of resources can be a particularly important function of a unified CZM program. It can identify in advance the most valuable coastal habitats, ecological functions, tourist attractions, and places subject to natural hazards damage. This information would be useful in the review and assessment of coastal development projects and in identifying candidate sites for coastal parks and protected areas. A survey would also locate highly polluted waters and degraded resources that need rehabilitation. Another benefit would be the location of optimal sites for future development activities.

An important survey job is identifying, evaluating, and delineating Special Habitats such as mangroves, coral reefs, dune fields, seagrass meadows, and other habitats. This includes all three categories: (i) generic categories of habitat that are always off limits for development (e.g. mangrove, coral reef, etc.); (ii) specific "environmentally sensitive areas" to be "red-flagged" for regulatory protection, through the project review process; and (iii) Special Habitats or areas to be recommended for parks, reserves, sanctuaries, refuges, or other protected area status (see Section 6.7).

The occasion may occur when the CZM professional either conducts or (more likely) supervises baseline survey work.

The design of baseline studies is most important to ensure that the correct parameters are selected and the methods are appropriate. Also, it is essential that the baseline can be duplicated during future monitoring of the sites (see Section 6.10.4).

It has become popular to enlist volunteers, professionals and non-professionals, for straightforward survey work; for example, the Reef Check program of 1997, organized by Gregor Hodgson enlisted hundreds of volunteers. Earthwatch also organizes volunteers for survey work. Such efforts must be closely supervised by professionals.

9.7 *As research coordinator*

In the implementation phase, key roles for natural and social scientists are to assist the CZM team in translating information from monitoring programs and assessing the efficacy of new measures. Some good advice to scientists involved in planning coastal management programs appeared in a United Nations (UN) report in 1996 [58]. Such scientists were advised to design their research by ". . . preparing concise statements of objectives for research and monitoring, clearly defining what is to be measured and why, and in identifying methodologies, facilities and personnel needed for the studies to be cost effective and successful". That is, for each priority issue to be addressed, scientists should work with CZM managers to formulate specific questions that are to be resolved through subsequent scientific investigations. The UN report lists eight major opportunities for scientific input [58]:

1 Environmental Impact Assessment (EIA);
2 Resource surveys;
3 Simulation modelling;
4 Economic assessment and valuation;
5 Legal and institutional analyses;
6 Social and cultural analyses;
7 Management methodologies;
8 Public education.

In project assessments scientists have important roles to play. In particular, they should evaluate the relevance, reliability, and cost effectiveness of scientific information generated by research and monitoring and advise on the suitability of control data. Scientists should also provide estimates of the extent to which observed changes in managed environments and practises are attributable to CZM measures as opposed to other factors [58].

The UN report notes that scientists can help bring together the information required by managers and politicians [58]. But often the reward system for scientists encourages them to concentrate on research which is not always relevant to management, a real problem for CZM staff.

9.8 *As environmental assessor*

A major function of CZM is to conduct EA of potential impacts of development. The EA process is the mechanism by which ecological and other environmental consequences of proposed development are estimated and recommendations provided to decision makers to reduce or avoid impacts (for details see Section 6.1). It is expected that social and economic impact will be included under EIAs. Training in the field is available, much of it supported by foreign aid.

9.9 *As mediator*

A major role of coastal management is to identify conflicts over coastal land and coastal renewable resources and to find ways to allocate and manage uses for the optimum long-term benefit of the nation by using a multiple use format. Methods for resolving conflicts include: fact finding and executive decision, study commissions, bargaining sessions, informal negotiation, facilitated discussion, formal mediation, administrative or public hearings, and adjudication [20]. Any method can be included in a CZM management framework (see Section 6.11).

9.10 *As linguist*

For advancement in any aspect of CZM, it is necessary to know the English language. In order to talk to people from other countries and to read essential literature, you will have to speak, read, and write English. To prove that you know English well you may have to take a test called TOEFL (Test of English as a Foreign Language; cost of test US$60) and get a score of at least 550. The TOEFL test is used by many organizations and universities in North America, but primarily for university admission. Tests include listening comprehension, structure and written comprehension, and reading comprehension. England, Australia, and New Zealand have a different test called IELTS (cost US$100).

9.11 *As student*

Most opportunities to gain specialized skills are at the masters or doctoral degree level (there are exceptions like Southern Cross University; see below). A graduate degree is usually required for advanced positions in CZM. But much of what you need for your job in CZM you can learn on your own, through informal study and experience. Peter Saenger [59] emphasizes that CZM is an "applied science" and therefore it must "integrate professional practise with academic theory".

Government staff in developing countries can find opportunities for subsidized degree or diploma programs in marine science, resources management, law, economics, or other appropriate CZM subject. Such opportunities abroad are provided by international assistance programs and require a foreign language capability (usually English). Students from developed countries who do study abroad usually have higher status on their return [30]. Financing your CZM education may be easier at the graduate level than at the undergraduate level because of scholarships and greater donor interest.

The following are some basic subjects that are relevant to CZM and that are variously offered at the undergraduate or graduate level, or are appropriate for self-study: marine biology, marine botany, taxonomy, geology, oceanography, chemistry/biochemistry, aquatic ecology, hydrology/hydraulics, statistics, meteorology, fisheries biology/management, computer technique, ornithology, mammology, planning principles, public health/toxicology, and cartography/drafting. Also, review courses in subjects like agriculture/forestry, law, government administration, civil/coastal engineering, and economics will be helpful. But CZM also requires *interdisciplinary* training because of the extent of coordinative work that is required (see Chapter 7) from national to local levels [60].

Each university will have its own particular curriculum of CZM subjects, such as pollution, marine technology, law of the sea, seaport management, marine tourism, aquaculture, marine economics, coastal zone law, and fish stock assessment. Most schools will incorporate active field study in the program whereby students visit sites and observe, measure, and report on issues and natural phenomena and derive solutions to CZM problems.

L. Jodice (Oregon State University, USA) provides the following advice to graduate students in CZM: Students should be encouraged to establish a better peer-like relationship with professors as well as other graduate students. It is also important when applying to a school to inquire about the particular research interests of their prospective professors. Also, in applying they should make it clear that they are interested in management as well as research. Once in school, they should be open and outgoing in communication with professors and this attitude will also help their job prospects later.

Several schools have specific CZM curricula. Others have broader marine oriented interdisciplinary programs which include CZM. In some schools, students and professors collaborate in creating customized training programs for each student. Following is a list of some universities that offer specialized curricula in CZM leading to Bachelor, "Professional", Masters, or PhD degrees.

HIGHER DEGREES
(Masters or PhD in Coastal Zone Management)

1 *Bogor Agricultural University*
Center for Coastal and Marine Resources Studies
Gd. Marine Center Lt. 4, Fakultas Perikanan IPB
Bogor, Indonesia

Contact: Dr Rokmin Dahuri Tel.:/Fax: 62-0251-624-185
E-mail: r-dahuri@indo.net.id

Program: Two-year Masters Degree in Coastal Zone Management. The course includes natural and social systems, policy, technologies, law, GIS, economics, and management methods. Thesis required. Lectures in Bahasa Indonesia language.

2 *Bournemouth University*
Department of Conservation Sciences
Talbot Campus
Fern Barrow, Poole, Dorset
BH12 5BB, United Kingdom

Contact: Faculty. Tel.: +44 (0) 1202 595178; Fax: 1202 595255

Program: Offers a one-year MSc in Coastal Zone Management (since 1992). Curriculum provides a good variety of practical management oriented CZM courses with lengthy off campus working experience and research topic.

3 *Duke University*
Nicholas School of the Environment
Duke Marine Laboratory
135 Duke Marine Lab Road
Beaufort, North Carolina 28516 USA

Contact: Dr Michael Orbach Tel.: 919-504-7605
E-mail: mko@mail.duke.edu

Program: Masters of Environmental Management degree. Provides a scientifically rigorous understanding of coastal environments and related policies and social aspects. Purpose is to prepare students for jobs in coastal policy and management, research, consulting firms, non-government organizations (NGOs), and government agencies. Good preparation for PhD study.

4 *Oregon State University*
College of Oceanic and Atmospheric Sciences
Marine Resources Management Program
Oceanography Admin Bldg 104
Corvallis, OR 97331 USA

Contact: Dr James E. Good. Tel.: 503-737-5118
E-mail: goodj@ccmail.orst.edu

Program: Masters Degree in Marine Resource Management. Course designed to train professionals for careers in wise development and management of marine and coastal resources; core courses in basic oceanography complemented with resource management, economics, policy, communications, geography, land use planning, and special courses in coastal and ocean management (degree program since 1974; 13 faculty members).

5 *University of Newcastle-upon-Tyne*
Centre for Tropical Coastal Management Studies
Department of Marine Sciences & Coastal Management
Newcastle-upon-Tyne
NE1 7RU, United Kingdom

Contact: Dr A. J. Edwards. Tel.: +44 (0) 191-222-6659
E-mail: a.j.edwards@newcastle.ac.uk

Program: Special one-year Masters Degree program in Coastal Zone Management (started in 1987); a broadly based course incorporating a multi-disciplinary approach to coastal management that includes environmental assessment, socio-economics, and resource management— strong emphasis on case studies in the Central Caribbean, South-East Asia, and Indian Ocean where faculty have worked extensively during the last 10 years. Solid base in ecological sciences.

6 *University of Technology, Sydney*
Institute for Coastal Resource Management
Westbourne St., Gore Hill
NSW 2065 Australia

Contact: Dr Kenneth Brown. Tel.: 61-2-330-4042
E-mail: Kenneth.Brown@uts.edu.au

Program: Offers MSc, PhD, and Diploma courses in CZM (from 1992). Broad curriculum begins with basic sciences and then enters into management with a variety of key CZM subjects.

HIGHER DEGREES
(Masters or PhD within a wider marine curriculum that includes Coastal Zone Management)

7 *Dalhousie University*
Marine Affairs Program
124 Seymour St.
Halifax, Nova Scotia
Canada B3H 3J5

Contact: Dr Aldo Chiracop. Tel.: 902-494-3555
E-mail: patricia.roberts@dal.c.
Web: http://www.dal.c./mmm

Program: Offers a one-year Masters in Marine Management. A broad
curriculum provides a mix of marine and coastal subjects, including CZM,
GIS, communities, EA, law, oceanography, fisheries, tourism, resource
reserves. Core course and electives. International focus.

8 *Nova Southeastern University*
Oceanographic Center
Institute of Marine and Coastal Studies
8000 N. Ocean Drive
Dania, FL 33004 USA

Contact: Dr Richard E. Dodge. Tel.: 954-920-1909
E-mail: dodge@ocean.nova.edu

Program: Offers a Masters combining Coastal Zone Management, Marine
Environmental Sciences, and Marine Biology—science orientation ($1\frac{3}{4}$
years).

9 *University of North Carolina at Chapel Hill*
Department of City and Regional Planning
CB# 3140, New East Building
Chapel Hill, North Carolina, 27599–3140 USA

Contact: David J. Brower. Tel.: 919-962-4775
E-mail: derp@mhs.unc.edu.

Program: Coastal Planning/Management specialization within Masters
of Regional Planning program. Faculty expertise includes growth
management, hazard mitigation, coastal law and policy, environmental
processes, and carrying capacity analysis; core coastal environment course
taught in Marine Science Department.

Also PhD in Regional Planning, CZM emphasis.

Contact: Dr D. Godschalk. Tel.: 919-962-5012
E-mail: dgod.dcrp@mhs.unc.edu

10 *University of Rhode Island*
Marine Affairs Program
Narragansett Bay Campus
Narragansett, Rhode Island 02882 USA

Contact: Dr Lawrence Juda. Tel.: 401-792-4041
E-mail: Ljuda@uriacc.uri.edu
Web site: www.uri.edu/artsci/maf

Program: URI offers both 2-year and 1-year (mostly for mid-career persons) Master of Marine Affairs degrees. Areas of specialization include fisheries, law, transportation, ports, policy, and coastal management. Also Bachelor degree programs.

Opportunities to interact with Coastal Resources Center with international CZM program.

Contact: Stephen Olsen. Tel.: 401-874-6224
E-mail: olsenuri@gsosun1.gso.uri.edu

11 *University of Washington*
School of Marine Affairs, HF-05
Seattle, Washington 98195 USA

Contact: Dr Marc J. Hershman. Tel.: 206-685-2469
E-mail: hershj@u.washington.edu

Program: The School offers a broad two-year Master of Marine Affairs program (themes: Coastal Zone Management, Marine Environmental Protection, Marine Policy, Port and Marine Transportation Management, and Fisheries Management.)

12 *University of the West Indies*
Faculty of Natural Sciences
Department of Zoology
St. Augustine
Republic of Trinidad and Tobago

Contact: Dr Peter Bacon Tel.: 809-663-2060, 663-2007, Exts. 2094, 3275.

Comment: Dr Bacon, Professor of Zoology, coordinates a course in Coastal

Management. He is senior author of the workbook *Practical Exercises in Coastal Management for Tropical Islands*.

BACHELORS DEGREE
(Bachelors in Coastal Zone Mangement)

13 *Deakin University*
662 Blackburn Road
Clayton, Victoria 3217
Australia

Contact: Dr Geoff Westcott. Tel.: 61-3-9244-7436
E-mail: wescott@deakin.edu.au

Program: Bachelor of Applied Science (coastal) with courses in ecology, EA, GIS, policy, land use, geology, ecology, planning, marine biology. Offers field work, work experience, laboratory work, study projects, and requires a "dissertation". Also, offers a "graduate diploma" course for mid-career training.

14 *Southern Cross University*
Centre for Coastal Management
PO Box 157
Lismore NSW 2480
Australia

Contact: Dr Peter Saenger. Tel.: 61-02-6620-3631
E-mail: psaenger@scu.edu.au

Program: Offers a three-year Bachelor of Applied Science degree in "Coastal Management" (since 1987); complete curriculum covering all important CZM topics; good facilities for both lab and field work; strong faculty.

Note: See also University of Rhode Island at 10 above.

SPECIAL DOCTORATE
(Universities without a prescribed curriculum in CZM but whose faculty have the experience to assist the student in organizing a PhD program with focus on CZM issues)

15 *University of Delaware*
College of Marine Studies
Newark, Delaware 19716 USA

Contact: Dr Robert Knecht. Tel.: 302-451-2336
E-mail: robert.knecht@mvs.udel.edu

16 *University of Wales*
Department of Maritime Studies
Cardiff, Wales CF1, United Kingdom

Contact: Dr R. Bollinger. Tel.: +44 (0)1222 87400

In addition some universities offer special summer institutes relevant to CZM, e.g. University of Rhode Island and Duke University (contact as above).

Further information is available from: (i) S.M. Vallejo, United Nations, NY. Tel.: 212-963-3935; E-mail: vallejo@un.org and (ii) C. Casullo, UNESCO, Room Mi.32, 1 rue Miollis, 75732 Paris, Cedex 15, France, Tel.: +33-45-68-30-08.

References

1 Clark, J.R. (1996) *Coastal Zone Management Handbook*, CRC/Lewis Publishers, New York. 704 pp.

2 Easterbrook, G. (1995) *A Moment on The Earth*, Penguin Books, New York. 745 pp.

3 Chua, T.E. & White, A.T. (1988) Policy recommendations for coastal area management in the ASEAN region, *ICLARM Contributions*, no. 544, International Center for Living Aquatic Resources Management, pp. 5–7.

4 Sorensen, J. (1997) National and international efforts at integrated coastal management: definitions achievements, and lessons, *Coastal Management*, Vol. 25, No.2, pp. 3–42.

5 Siddiqi, N.A. (1997) Management of resources in the Sundarbans Mangroves of Bangladesh, *Intercoast Network*, University of Rhode Island (USA), March/1997, pp. 22–23.

6 Clark, J.R. (1983) *Coastal Ecosystems Management: A Technical Manual for the Conservation of Coastal Zone Resources*, Robert E. Krieger Publishing Co., Melbourne, Florida (1st edn by John Wiley–Interscience, New York, 1977). 928 pp.

7 Ormond, A., Shepherd, A.D., Price, A. & Pitts, R. (1985) *Management of Red Sea Coastal Resources: Recommendations for Protected Areas*, Report No. 5. Saudi Arabia Marine Conservation Programme, International Union for Conservation of Nature, Gland, Switzerland. 113 pp.

8 WRI (1986) *World Resources 1986*, World Resources Institute, Washington DC. 353 pp.

9 Clark, J.R., Banta, J.S. & Zinn, J.A. (1980) *Coastal Environmental Management, Guidelines for Conservation of Resources and Protection against Storm Hazards*, US Government Printing Office, Washington DC. 161 pp.

10 Brown, L.R. (1985) *State of the World*, W.W. Norton & Co, New York. 301 pp.

11 Article in *The Straights Times* (Singapore), August 5 (1997)

12 NACRF n.d. *Community Action Program for Water Pollution Control*, National Association of Counties Research Foundation, Washington DC.

13 Tomasik, T., Mah, A., Nonti, A. & Moosa, M. (1997) *The Ecology of the Indonesian Seas* (Parts 1 & 2), Periplus edns, Singapore. 1388 pp.

14 Cambers, G. (ed.) (1997) *Managing Beach Resources in the Smaller Caribbean Islands*, pp. 63–68. University of Puerto Rico/Sea Grant.

15 Chua, T.E. & Charles, J.R. (1984) *Coastal Resources of East Coast Peninsular Malaysia*, p. 306. Penerbit University of Sains Malaysia.

16 Charlier, R.H. & DeCroo, D. (1991) *Coastal Erosion. Response and Management*, Haecon N.V., Ghent, Belgium, IWK 133/91.04462.

17 Saenger, P. (1989) *Environmental impacts of coastal tourism: an overview and guide to relevant literature*, Informal Paper, Centre for Coastal Management, University of New England, NSW, Australia. 17 pp.

18 Dunkel, D.R. (1984) Tourism and the environment: a review of the literature and Issues, *Environmental Sociology*, no. 37.

19 Tomascik, T. (1992) *Environmental Management Guidelines for Coral Reef Ecosystems*, State Ministry for Population and Environment (Kependudukan dan Lingkungan Hidup), Jakarta. 116 pp.

20 Sorensen, J.C. & McCreary, S.T. (1990) *Institutional Arrangements for Managing Coastal Resources and Environments*, Coastal Management Publication No. 1 [Rev.], NPS/US AID Series, National Park Service, Office of International Affairs, Washington DC. 194 pp.

21 Geoghegan, T., Jackson, I., Putney, A. & Renard, Y. (1984) *Environmental Guidelines for Development in the Lesser Antilles.*, Eastern Caribbean Natural Areas Management Program, St. Croix, U.S.V. I. 44 pp.

22 Marcos, I. (1983) *National Environmental Enhancement Program*, National Environmental Protection Council, Imelda Marcos, Chairperson, Manila. 137 pp.

23 Boelart-Suominen, S. & Cullinan, C. (1994) *Legal and Institutional Aspects of Integrated Coastal Area Management in National Legislation*, United Nations, Development Law Service, Rome. 118 pp.

24 Clark, J.R. (1992) *Integrated Management of Coastal Zones*, Fisheries Tech. Paper no. 327. United Nations/FAO, Rome. 167 pp.

25 Hodgson, G. & Dixon, J.A. (1988) *Logging Versus Fisheries and Tourism in Palawan*, Occasional Paper no. 7, Environment and Policy Institute, East–West Center, Hawaii. 95 pp.

26 Salm, R.V. &. Clark, J.R. (1989) *Marine and Coastal Protected Areas: A Guide for Planners and Managers*, 2nd edn. International Union for the Conservation of Nature, Gland, Switzerland. 302 pp.

27 News item in *World Conservation 2/96*, International Union for Conservation of Nature, Gland, Switzerland. 121 pp.

28 Kana, T.W. (1991) Treating the coast as a dynamic system. In *The Status of Integrated Coastal Zone Management: A Global Assessment*, Clark, J.R. (ed.), pp. 60–61. CAMPNET. University Of Miami/RSMAS, Miami, Florida.

29 Clark, J.R. (ed.) (1991) *The Status of Integrated Coastal Zone Management: A Global Assessment*, CAMPNET, University Of Miami/RSMAS, Miami, Florida. 118 pp.

30 Robinson, A.H. (1988) Staff training for coral reef and other marine area management. In *Coral Reef Management Handbook*, pp. 147–162. Kenchington, R.A. & Hudson, B.E.T. (eds.) UNESCO Regional Office for Science and Technology for South-East Asia, Jakarta.

31 Olsen, S. (1994) Personal communication.

32 Jernelov, A. & Marinov, U. (1990) *An Approach to Environmental Impact Assessment for Projects Affecting the Marine Environment*, UN/Environmental Program, Regional Seas Reports and Studies, No. 122. 37 pp.

33 Carpenter, R.A. & Maragos, J.E. (1989) *How to Assess Environmental Impacts on Tropical Islands and Coastal Areas.* East–West Center, Honolulu, Hawaii. 345 pp.

34 Day, J.W. Jr, Martin, J.F., Cardoch, L. & Templet, P.H. (1997) System functioning as a basis for sustainable management of deltaic ecosystems, *Coastal Management*, Vol. 25 No. 2, pp. 115–153.

35 Pabla, H.S., Pandey, S. & Badola, R. (1993) *Guidelines for Ecodevelopment Planning*, UNDP/FAO Project, Wildlife Institute of India, Dehradun, India. 40 pp.

36 Clark, J.R. (ed.) (1991) *Carrying Capacity, A Status Report on Marine and Coastal Parks and Reserves*, University of Miami, RSMAS, Miami, Florida. 73 pp.

37 Chua, T.E. & Scurd, L.F. (1992) *Integrative Framework and Methods for Coastal Area Management.* ICLARM, Manila. 169 pp.

38 IUCN (1990) *Caring for the World: A Strategy for Sustainabilty*, International Union for Conservation of Nature, Gland, Switzerland. Second Draft, June, 1990. 135 pp.

39 Heyman, A.M. (1986) *Inventory of Caribbean Marine and Coastal Protected Areas*, Organization of American States, Washington DC. 146 pp.

40 Kenchington, R.A. (1990) *Managing Marine Environments*, Taylor and Francis, New York.

41 Clark, J.R. (1988) *Rehabilitation of coral reef habitats*. Rept. of a Science Workshop held at St. John, USVI, December, 1987 (unpublished), University of Miami/RSMAS and US National Park Service. 16 pp.

42 Renard, Y. (1996) Sharing the benefits of coastal conservation. *World Conservation* 2/96, International Union for Conservation of Nature Gland, Switzerland. pp. 21–23.

43 Pheng, K.S. & Kam, W.P. (1989) Geographic information systems in resource assessment and planning, *Tropical Coastal Area Management*, Vol 4, No. 2, International Center for Living Aquatic Resources Management, Manila.

44 Butler, M.J.A., LeBlanc, C., Belbin, J.A. & MacNeill, J.L. (1987) *Marine Resource Mapping: An Introductory Manual*, United Nations, FAO, Fisheries Technical Paper No. 274. 256 pp.

45 Crawford, B.R., Cobb, J.S. & Ming, C.L. (eds) (1996) *Educating Coastal Managers*. Proceedings of the Rhode Island Workshop. Coastal Research Center, University of Rhode Island. 170 pp.

46 Kenchington, R.A. & Crawford, D. (1993) On the meaning of integration in coastal zone management, *Ocean and Coastal Management*, Vol 21, Nos 1–3, pp. 109–127.

47 Renard, Y. (1986) Citizen Participation in Coastal Area Planning and Management. *CAMP Newsletter*, pp. 1–3. October 1986. National Park Service, Office of International Affairs, Washington DC.

48 Kelleher, G. (1996) Public participation on 'the Reef'. *World Conservation*, 2/96, pp. 21–23. International Union for Conservation of Nature, Gland, Switzerland, Gland, Switzerland. 19 pp.

49 White, A. Why public participation is important for marine protected areas, *CAMP Newsletter*, pp. 5–6. August (1987) National Park Service, Office of International Affairs, Washington DC.

50 Anonymous. (1996) Collaborative Management. *World Conservation*, 2/96, International Union for Conservation of Nature, Gland, Switzerland. 3 pp.

51 Clark, J.R. (1992) *Ujung Pandang Port Urgent Rehabilitation Project: Environmental Report*, Indonesia Department of Communications.

52 White, A.T. (1988b) The effect of community-managed marine reserves in the Philippines on their associated coral reef fish populations. *Asian Fisheries Science* Vol 1, No. 2, pp. 27–42.

53 White, A.T. & Calumpong, H. (1992) Saving Tubbataha Reefs and monitoring marine reserves in the Central Visayas, *Summary Field Report*, Earthwatch Expedition, Philippines, April–May 1992, unpublished.

54 Sofield, T.H.B. (1990) The impact of tourism development on traditional sociocultural values in the South Pacific: conflict, coexistence, and symbiosis. In *Proceedings of the 1990 Congress on Coastal and Marine Tourism*, Vol. II, pp. 49–66. Miller, M. L. & Auyong, J. (eds) National Coastal Resources Research and Development Institute, Newport, Oregon (USA).

55 Olsen, S. (1987) Sri Lanka completes its coastal zone management plan. *CAMP Newsletter*, National Park Service, Office of International Affairs, Washington DC. August, 1987.

56 CCD (1990) *Coastal Zone Management Plan*, Sri Lanka Coast Conservation Department, Colombo, Sri Lanka. 110 pp.

57 Olsen, S. (1993) Will integrated coastal management programs be sustainable: the constituency problem. *Ocean and Coastal Management*, Vol 21, Nos. 1–3, pp. 201–225.

References

58 GESAMP (1996) *The Contribution of Science to Integrated Coastal Management*, United Nations/FAO, Reports & Studies No. 61. 66 pp.

59 Saenger, P. (1994) Integrated coastal management education: the experience of the Centre for Coastal Management. In *Recent Advances in Marine Science and Technology '94*, pp. 523–532. Pacon International and James Cook University.

60 Hay, J. (1995) Education, training, and networking in Coastal Zone Management, in K. Hotta and I.M. Dutton. In *Coastal Management in the Asia-Pacific Region: Issues and Approaches*, pp. 133–152. Japan International Marine Science and Technology Federation, Tokyo.

Glossary

artisanal economic activities based on personal skills and manual dexterity; generally "low tech" work, often resource oriented.

baseline study an inventory of a natural community or environment to provide a baseline —a measure of its condition at a point of time—often done to describe the status of biodiversity and abundance against which future change can be gauged (usually development driven).

buffer area a protective, often transitional, area of controlled use — in coastal management, a peripheral zone separating a developed area from a protected natural area.

carrying capacity the limit to the amount of life, or economic activity, that can be supported by an environment; the reasonable limits of human occupancy and/or resource use.

Coastal Zone a zone comprising coastal waters (including the lands thereunder) and the adjacent shorelands; the zone strongly influenced by both sea and land and including smaller near-coast islands, transitional and intertidal areas, wetlands (mangroves and marshes) and beaches.

Coastal Zone Management (CZM) a governmental process for achieving sustainable use of resources of the Coastal Zone whereby participation by all affected economic sectors, government agencies, and non-government organizations is involved; unified or integrated Coastal Zone Management when the management actions of the various stakeholders are formally unified and community participation is emphasized.

Co-management the process whereby authority for management is shared between communities and higher levels of government. Also "community based management" or "collaborative management".

Commons publicly owned areas of land or water, often managed by government as a public trust for the people; common property.

ecotourism tourist activity attracted to environmental resources and based, usually, on a conservation theme.

Environmental Impact Assessment (EIA) detailed prediction of the impact of a development project on environment and natural resources with recommendations as to acceptability of the project, need for minimizing/eliminating/offsetting adverse effects, and a management plan to accomplish these countermeasures; a generic term for all types of impact assessment is Environmental Assessment (EA).

estuary a semienclosed littoral basin (embayment) of the coast in which fresh river water entering at its head mixes with saline water entering from the ocean.

Exclusive Economic Zone the maritime zone adjacent to and extending 200 nautical miles beyond the baseline from which the territorial sea is measured—internationally authorized by the Third United Nations Conference on the Law of the Sea; the coastal state has sovereign rights to explore, exploit, conserve and manage the natural resources in this zone.

Geographic Information System (GIS) computer-assisted systems that can input, store, retrieve, analyze and display geographically referenced information and enhance the analysis and display of interpreted geographic data.

greenhouse effect heating of the Earth from the increase in the gases such as CO_2, methane, CFCs, etc., that make up the atmospheric envelope that surrounds the globe; term coined by the scientist Svante Arrhenius in the late 1800s.

ground truth ground level direct observations made to verify interpretations from remotely sensed data.

integrated Coastal Zone Management see **Coastal Zone Management**.

infrastructure usually the publicly constructed support system for a community including roads, electricity, communications, water, sewage, etc.

issues analysis the exploration, definition, and evaluation of the basic resource management issues to be faced in an ICZM program.

lagoon (i) a semienclosed littoral basin with limited fresh water input, high salinity, and restricted circulation; lagoons often lie behind sand-dunes, barrier islands, or other protective features; (ii) the shallow waters lying between a coral ridge and the shore.

littoral of or pertaining to the shore, especially of the sea; coastal.

Master Plan the operational CZM plan which defines rules, resources, conservation issues, performance standards, authorities, objectives, use rights (permitted uses), development restrictions, participation, coordination mechanisms, permit/EIA conditions, protected areas, setbacks, staff, training, etc.

mitigation the elimination, reduction or control of the adverse environmental impacts of a project, including countermeasures against negative environmental impacts of development.

multiple use the concept of providing for multiple activities for particular areas or resources by managing them for sustainable resource use.

nurture (or nurturing) area any place in the coastal zone where larval, juvenile, or young stages of aquatic life concentrate for feeding or refuge; also a "nursery area".

pollutant a contaminant that in a certain concentration or amount will adversely alter the physical, chemical, or biological properties of the environment — includes pathogens, heavy metals, carcinogens, oxygen-demanding materials, and all other harmful substances, including dredged spoil, solid waste, incinerator residue, sewage, garbage, sewage sludge, munitions, chemical wastes, biological materials, radioactive materials, and industrial, municipal, and agricultural wastes discharged into coastal waters.

primary waste treatment a process that removes material that floats or will settle in sewage, accomplished by using screens to catch the floating objects and settling tanks for heavy matter, and often including chlorination; removes of about 30% of BOD and less than half of metals and toxic organics.

protected area a natural area of land or water set aside by governmental action, as a right of ownership, to protect its resources from degradation.

Rapid Rural Assessment (RRA) a procedure for gathering and analyzing information about community socio-economic conditions preparatory to making development decisions; where community participation is a priority; also "Participatory Rural Assessment" (PRA) or "Rapid Coastal Assessment" (RCA).

red tide a massive "bloom" of dinoflagellate microscopic organisms that may produce neurotoxins such as paralytic shellfish poisoning (PSP) that infest marine organisms and humans that eat them; may kill fish and pollute the air with irritating substance; red or reddish brown discoloration of the sea.

remote sensing the acquisition and processing of information about a distant object or phenomenon without any physical contact; often done from satellites.

reserve (nature or resource reserve) an area designated for protection (and restoration) of environmental resources, as a right of governmental ownership, which requires limitation of exploitive use.

retreat a coastal land use strategy whereby structural development is withdrawn from the coast to a designated setback line farther inland.

sea level rise the increase in elevation of the sea caused by the Greenhouse Effect which results from heat expansion of the ocean waters and meltdown of the Polar ice caps; recognized by the writer Jules Verne nearly a century ago.

secondary treatment sewage treatment that follows primary treatment to consume organic part of wastes by bringing sewage and bacteria together in "trickling filters" or in "activated sludge processes"; may remove up to 90% of BOD by converting organics to inorganics (e.g., phosphate and nitrate).

sector a distinct component of the economic and political structure of a country, e.g., the manufacturing sector or the public sector.

setback a prescriptive linear space which is often specified in shoreline management programs, to separate development sites from natural areas or to remove structures inland away from the danger of sea storms or erosion.

shorelands the dry side of the coastal zone; low-lying areas that are affected by coastal waters through flooding, air borne salt, or other marine processes.

situation management coastal management programs that focus on particular problems or areas, rather than the whole coastal zone of a country; geo-specific or issue-specific coastal management programs.

Special Habitat an area of highly concentrated biological activity of a type that is especially valuable for maintaining biodiversity and/or resource productivity; an ecologically sensitive or critical area or habitat.

stakeholder a person (or entity) having a vested interest in decisions affecting the use and conservation of coastal resources.

storm surge a rise of sea elevation caused by water piling up against a coast under the force of strong onshore winds such as those accompanying a hurricane or other intense storm; reduced atmospheric pressure may contribute to rise.

Strategy Plan the first stage in coastal planning whereby the basic national strategy for ICZM is decided, including analysis of issues, needs, goals, objectives, and equities.

sustainable use practises that ensure the continuance of natural resource productivity and a high level of environmental quality, thereby providing for economic growth to meet the needs of the present without compromising the needs of future generations.

unified Coastal Zone Management see **Coastal Zone Management.**

wetlands low-lying vegetated areas that are flooded at a sufficient frequency to support vegetation adapted for life in saturated soils, including mangrove swamps, salt marshes, and other wet vegetated areas (often between low water and the yearly normal maximum flood water level).

Zone of Influence an area adjacent to the Coastal Zone which influences the condition of it's resources and for which a mechanism is created for coordination with a Coastal Zone Management program.

zoning a system of designating areas of land or water to be allocated to specific (often exclusive) uses; the division of a particular area into several zones, each of which is scheduled for a particular use or set of uses.

Unit Conversion Table

	Mulitply number of	by	to obtain equivalent number of
Length	Millimeters	0.03937	inches
	Centimeters	0.3937	inches
	Meters	39.3701	inches
		3.2808	feet
		1.0936	yards
		0.54681	fathoms
	Kilometers	0.62137	miles (land)
		0.53961	miles (UK sea)
		0.53996	miles, international nautical
Area	Square millimeters	0.00155	square inches
	Square centimeters	0.1550	square inches
	Square meters	10.7639	square feet
		1.19599	square yards
	Hectares	2.47105	acres
	Square kilometers	247.105	acres
		0.3861	square miles
Volume and capacity	Cubic centimeters	0.06102	cubic inches
	Liters	61.024	cubic inches
		0.0353	cubic feet
		0.2642	US gallons
		0.2200	UK gallons
	Cubic meters	35.3147	cubic feet
		1.30795	cubic yards
		264.172	US gallons
		219.969	UK gallons
		6.11025	UK bulk barrels
Weight (mass)	Grams	0.03527	ounces, avoirdupois
		0.03215	ounces, troy
	Kilograms	2.20462	pounds, avoirdupois
	Tonnes	2204.62	pounds, avoirdupois
		1.10231	short tons
		0.984207	long tons

Index